Higher Grade Biology Multiple Choice Tests

Higher Grade
BIOLOGY

Multiple Choice Tests

Team Co-ordinator
James Torrance

Writing Team
James Torrance
James Fullarton
Clare Marsh
James Simms
Caroline Stevenson

Diagrams by James Torrance

Hodder & Stoughton
LONDON SYDNEY AUCKLAND TORONTO

CLASS NO 574 TOR
ITEM NO 330005147x

THE LIBRARY
WESTMINSTER COLLEGE
OXFORD   OX2 9AT

Contents

Preface *vii*

1. Cell variety in relation to function *1*
2. Absorption and secretion of materials *3*
3. ATP and energy release *7*
4. Chemistry of respiration *9*
5. Absorption of light by photosynthetic pigments *12*
6. Chemistry of photosynthesis *13*
7. DNA and its replication *15*
8. RNA and protein synthesis *18*
9. Variety of proteins *20*
10. Viruses *22*
11. Cellular response in defence *24*
12. Meiosis *27*
13. Monohybrid cross *30*
14. Dihybrid cross *33*
15. Sex linkage *35*
16. Mutation *37*
17. Natural selection *39*
18. Selective breeding of animals and plants *42*
19. Loss of genetic diversity *45*
20. Genetic engineering *46*
21. Speciation *48*
22. Adaptive radiation *51*
23. Growth differences between plants and animals *52*
24. Growth patterns *57*
25. Genetic control *59*
26. Control of gene action *61*
27. Hormonal influences on growth – part 1 *63*
28. Hormonal influences on growth – part 2 *66*
29. Effects of chemicals on growth *68*
30. Effect of light on growth *70*
31. Physiological homeostasis *73*
32. Population dynamics – part 1 *76*
33. Population dynamics – part 2 *78*
34. Population dynamics – part 3 *79*
35. Maintaining a water balance – animals *81*
36. Maintaining a water balance – plants *83*
37. Obtaining food – animals *86*
38. Obtaining food – plants *89*
39. Coping with dangers – animals *91*
40. Coping with dangers – plants *92*
    Specimen Examination Paper *94*

Preface

This book is intended to act as a valuable resource to pupils and teachers by providing a bank of multiple choice items whose content adheres closely to the syllabus for SCE Higher Grade Biology to be examined in and after 1991.

Each test matches part of a syllabus sub-topic and contains a variety of types of multiple choice item, many testing *knowledge* and *understanding*, and some testing *problem-solving* skills.

The tests are followed by a 40-item specimen examination paper modelled in the style of Paper I of the new examination.

Test 1 Cell variety in relation to function

1 Which of the following is the basic unit of life (i.e. the smallest unit that can lead an independent existence)?
   A molecule   B tissue
   C cell   D nucleus

Items **2** and **3** refer to the following diagram of a unicellular organism called *Chlamydomonas* which is able to swim about in its natural habitat (stagnant water).

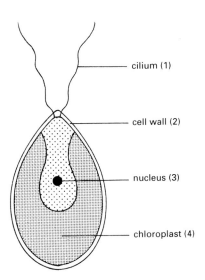

2 *Chlamydomonas* is
   A an animal.   B a bacterium.
   C a fungus.   D a green plant.

3 In the following table, which set of numbers correctly relates the labelled structures in the diagram to the functions that they perform?

	function			
	control of cell activities	movement	photo-synthesis	protection
A	1	3	4	2
B	3	1	2	4
C	4	2	3	1
D	3	1	4	2

Items **4** and **5** refer to the following diagram of the organism *Euglena*.

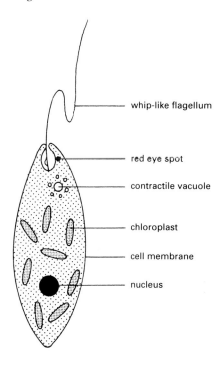

4 *Euglena* could be classified as an animal due to the
   A absence of a cell wall.
   B presence of a nucleus.
   C absence of motile cilia.
   D presence of a cell membrane.

5 *Euglena* could be classified as a plant due to the presence of a
   A cell membrane.   B chloroplast.
   C contractile vacuole. D flagellum.

6 The function of goblet cells present in ciliated epithelium in the human trachea is to
   A release carbon dioxide.
   B secrete mucus.
   C sweep microbes away.
   D absorb oxygen.

Items **7** and **8** refer to the diagrams of human cells shown overleaf (not drawn to scale).

7 Which cell's structure is suited to the function of transmitting nerve impulses?

8 Which cell's structure is suited to the function of oxygen uptake?

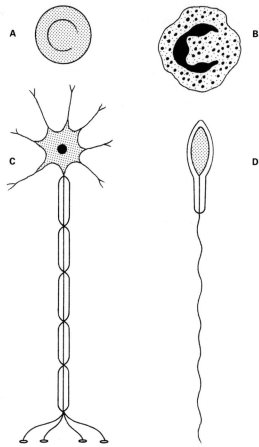

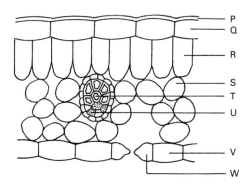

9 Epithelium from the trachea and from the internal lining of the cheek *both* possess
   A squamous cells.   B columnar cells.
   C ciliated cells.   D goblet cells.

10 Which of the following tubes present in the human body is lined with ciliated epithelium?
   A oviduct   B artery
   C intestine   D vein

11 The diagram below shows a small piece of human tissue which consists of fibrous cords. These are neither elastic nor pliable.
   This tissue is called a
   A ligament, and it attaches muscle to bone.
   B ligament, and it attaches bone to bone.
   C tendon, and it attaches muscle to bone.
   D tendon, and it attaches bone to bone.

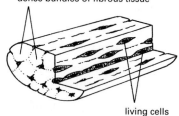

12 Which of the following are *both* composed of cells whose structure suits their function of protection?
   A P and Q   B Q and V
   C P and V   D P and W

13 Photosynthesis occurs in
   A Q, R and S.   B R, S and W.
   C P, Q and R.   D R, S and V.

14 Which cells are structurally adapted to suit their function of controlling stomatal size?
   A T   B U   C V   D W

15 Which tissue contains cells structurally adapted to translocate soluble carbohydrates to other parts of the plant?
   A R   B S   C T   D U

16 The diagram below shows a type of plant cell.

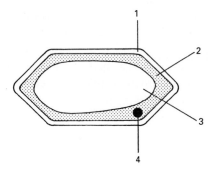

Compared with this cell, a mature xylem vessel would possess
   A 1 only.   B 1 and 2 only.
   C 1, 2 and 3 only.   D 1, 2, 3 and 4.

17 Which of the following are *both* structurally suited to perform the functions of water transport and support in a plant?
A sieve tube and companion cell
B xylem vessel and companion cell
C tracheid and xylem vessel
D sieve tube and tracheid

Items **18**, **19** and **20** refer to the following possible answers.
A sieve tube B root hair
C xylem vessel D tracheid

18 Which cell type possesses a nucleus and presents a large surface area for water absorption?

19 Which cell type lacks a nucleus but possesses cytoplasmic strands in contact with similar neighbouring cells?

20 Which cell type is non-living and has thickened end-walls that overlap with other similar cells?

Test 2 Absorption and secretion of materials

Items **1**, **2** and **3** refer to the following diagram which shows ways in which molecules may move into and out of a respiring animal cell.

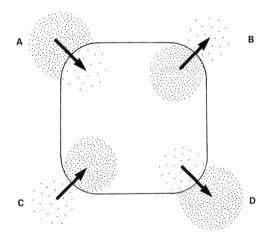

1 Which of these could be diffusion of carbon dioxide molecules?

2 Which of these could be active transport of sodium ions out of the cell?

3 Which of these could be active uptake of potassium ions?

4 Osmosis is the passage through a selectively permeable membrane of
A water from a region of higher solute concentration to a region of lower solute concentration.
B solute from a region of lower water concentration to a region of higher water concentration.
C water from a region of lower solute concentration to a region of higher solute concentration.
D solute from a region of higher water concentration to a region of lower water concentration.

Items **5** and **6** refer to the experiment shown in the following diagram.

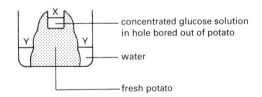

5 After a few days, which of the following will have occurred?
A a rise in level X and a drop in level Y
B a drop in level X and a drop in level Y
C a rise in level X and a rise in level Y
D a drop in level X and a rise in level Y

6 Which of the diagrams overleaf shows a suitable control for the above experiment?

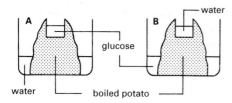
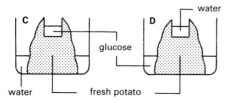

7 Which of the following are *all* present in a cell wall?
  A cellulose fibres, calcium pectate, phospholipids and enzymes
  B water-filled spaces, enzymes, cellulose fibres and calcium pectate
  C calcium pectate, plasma membranes, water-filled spaces and enzymes
  D enzymes, cellulose fibres, phospholipids and water-filled spaces

8 The following diagram shows the fluid-mosaic model of the structure of a cell membrane.

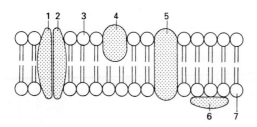

Which pair of structures numbered in the diagram are correctly identified in the following table?

	protein	phospholipid
A	1	7
B	2	4
C	3	7
D	5	6

9 Which of the following sucrose solutions has the highest water concentration?
  A 1.1 molar   B 0.8 molar
  C 0.5 molar   D 0.1 molar

10 The following diagram shows the appearance of a plant cell immersed in a solution which is isotonic to the cell's sap.

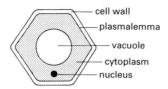

Which of the diagrams shown below most accurately represents the appearance of this cell after immersion in a hypertonic solution?

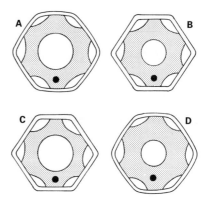

Items **11** and **12** refer to the following graph of the results from an experiment where each of four potato cylinders was immersed in a different chemical for a time, and then placed in water.

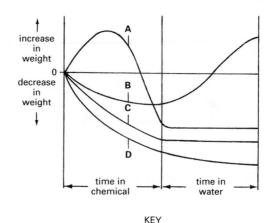

KEY
A = cylinder **A** in chemical **A**
B = cylinder **B** in chemical **B**
C = cylinder **C** in chemical **C**
D = cylinder **D** in chemical **D**

11 Which chemical was *not* toxic to the selectively permeable membranes of potato cells?

12 Deplasmolysis is the opposite process from plasmolysis. In which cylinder did deplasmolysis occur?

Items **13** and **14** refer to the following diagram of four plant cells.

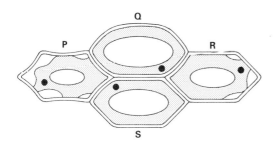

13 No wall pressure would exist in cells
 A P and Q.   B Q and S.
 C P and R.   D R and S.

14 If the cells remained in contact as shown, then water would pass by osmosis from *both*
 A R to Q and Q to P.
 B Q to S and R to Q.
 C P to Q and R to S.
 D Q to P and Q to R.

15 In an experiment, groups of potato discs were weighed and then each group was immersed in one of a series of sucrose solutions. After two hours each group was reweighed and its percentage gain or loss in weight was calculated.
 The following graph shows the results plotted as points.

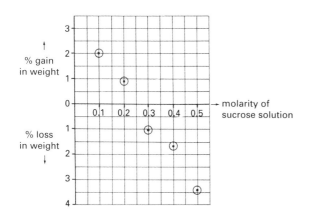

From these results it can be concluded that the water concentration of the potato cell sap is approximately equivalent to that of a sucrose solution of molarity
 A 0.15   B 0.25   C 0.35   D 0.45

16 The diagram below shows the results of an analysis of cell sap from a marine plant, and of the surrounding sea water.

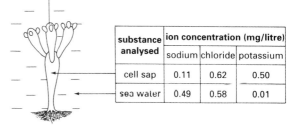

From this data it can be concluded that this plant
 A accumulates all three types of ion from sea water.
 B holds chloride ions at a concentration lower than sea water.
 C selects and internally accumulates sodium ions only.
 D can discriminate between sodium and potassium ions.

17 The table below shows the outcome of an investigation into the uptake of bromide ions by a plant.

time from start of experiment in minutes	units of bromide ions taken up by plant tissue under the following conditions		
	sugar absent, oxygen present	sugar present, oxygen absent	sugar and oxygen present
0	0	0	0
30	0	30	100
60	0	50	150
90	0	70	180
120	0	70	200

These results indicate that uptake of bromide ions
 A is an active process requiring energy.
 B occurs during aerobic respiration only.
 C requires a temperature suitable for enzymes to act.
 D stops in the absence of oxygen.

Items **18** and **19** refer to the following graphs.

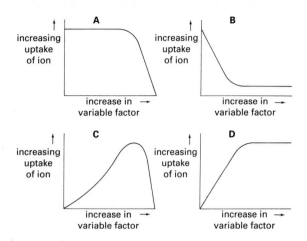

18 Which graph represents rate of ion uptake by a living cell in response to increasing oxygen concentration?

19 Which graph represents rate of ion uptake by a living cell in response to increasing temperature?

20 The following graph shows the changes in ionic concentrations of culture solutions in which barley roots were grown for two days.

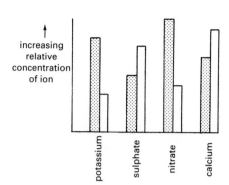

Which of the following ions were *both* taken up by the plant?
A calcium and sulphate
B sulphate and potassium
C potassium and nitrate
D nitrate and calcium

Test 3 ATP and energy release

1 Which of the following diagrams best represents the structure of a molecule of ATP (adenosine triphosphate)?

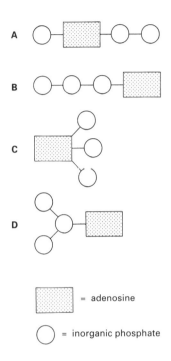

2 Which of the following equations represents the regeneration of ATP from its components?
A ADP + Pi $\xrightarrow{\text{energy taken in}}$ ATP
B ADP + Pi + Pi $\xrightarrow{\text{energy taken in}}$ ATP
C ADP + Pi $\xrightarrow{\text{energy released}}$ ATP
D ADP + Pi + Pi $\xrightarrow{\text{energy released}}$ ATP

3 The regeneration of ATP from its components is called
A oxidation.　　　　B metabolism.
C respiration.　　　D phosphorylation.

4 As part of an investigation into the effect of different solutions on fresh muscle tissue, 12 drops of ATP were added to a strand of fresh muscle of initial length 50 mm. After a few minutes its length was found to be 42 mm.
   Which of the following correctly summarises the experiment?

	% difference in length of muscle strand	reason for change
A	8	contraction of muscle fibres
B	8	relaxation of muscle fibres
C	16	contraction of muscle fibres
D	16	relaxation of muscle fibres

Items 5 and 6 refer to the following possible answers.
A removal of hydrogen ions from substrate and release of energy
B addition of hydrogen ions to substrate and consumption of energy
C removal of hydrogen ions from substrate and consumption of energy
D addition of hydrogen ions to substrate and release of energy

5 Which statement describes the process of oxidation?

6 Which statement describes the process of reduction?

Items 7 and 8 refer to the following diagram illustrating tissue respiration.

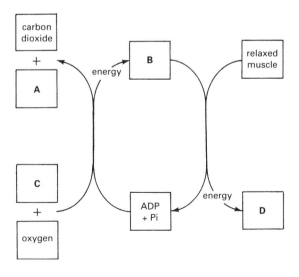

7 Which box represents the correct position of ATP in this scheme?

8 Which box represents the correct position of glucose in this scheme?

9 A muscle will contract in the presence of
   A ADP at 60°C.   B ATP at 60°C.
   C ADP at 30°C.   D ATP at 30°C.

10 Which of the following is an example of anabolism (the synthetic reactions of metabolism which require energy from ATP)?
   A deamination of amino acids to form urea
   B digestion of starch by enzymes
   C formation of protein from amino acids
   D oxidation of glucose during respiration

Test 4 Chemistry of respiration

Items **1**, **2**, **3** and **4** refer to the following diagram which shows a simplified summary of aerobic respiration. (Each intermediate compound is represented by a box containing the number of carbon atoms present in one molecule of that compound.)

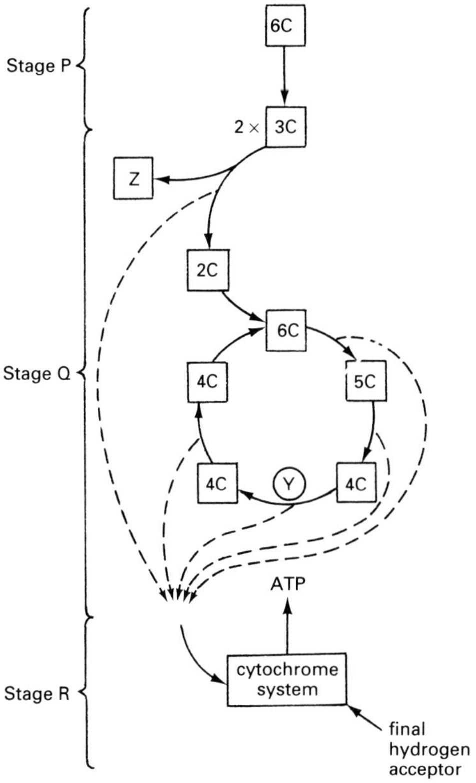

	cristae	cytoplasm	central cavity of mitochondrion
A	R	P	Q
B	Q	P	R
C	P	R	Q
D	R	Q	P

2 Substance Z is
   A water.
   B lactic acid.
   C carbon dioxide.
   D adenosine diphosphate.

3 The enzyme which controls reaction Y is a
   A decarboxylase.   B phosphorylase.
   C peroxidase.   D dehydrogenase.

4 The final hydrogen acceptor in the cytochrome system is
   A water.   B oxygen.
   C coenzyme.   D ADP.

5 During aerobic respiration of one molecule of glucose, most ATP is synthesised during
   A glycolysis.
   B Krebs cycle.
   C hydrogen transfer along the cytochrome system.
   D breakdown of pyruvic acid to 2-carbon compound.

6 How many molecules of ATP are synthesised as a result of the complete oxidation of one molecule of glucose?
   A 2   B 4   C 36   D 38

7 Which of the following are *both* required by a living cell for glycolysis to occur?
   A glucose and oxygen
   B ATP and glucose
   C oxygen and ATP
   D pyruvic acid and oxygen

8 The enzymes required for the Krebs cycle in a plant cell are located in the
   A cytoplasmic fluid surrounding each mitochondrion.
   B cristae of each mitochondrion.
   C outer membrane of each mitochondrion.
   D central matrix of each mitochondrion.

1 Which row in the following table indicates the correct locations at which stages P, Q and R occur in a cell?

Items **9** and **10** refer to the following diagram of a mitochondrion.

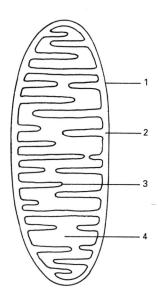

9 Stalked particles would be found at position
   A 1.   **B** 2.   **C** 3.   **D** 4.

10 Which of the following is *least* likely to be found in region 4?
   A glucose   **B** ATP
   C citric acid   **D** ADP

11 Which of the following would possess *fewest* mitochondria per unit volume of cell?
   A motile sperm cell   **B** nerve cell
   C cheek epithelial cell **D** liver cell

12 To respire anaerobically, a yeast cell needs
   A alcohol.   **B** glucose.
   C lactic acid. **D** cytochrome.

Items **13** and **14** refer to the following possible answers.
A ethanol + CO_2 + ATP
B ethanol + ADP
C lactic acid + CO_2 + ADP
D lactic acid + ATP

13 Which answer correctly identifies the end products of anaerobic respiration in a water-logged root cell?

14 Which answer correctly identifies the end products of anaerobic respiration in mammalian muscle tissue?

15 The following table refers to four test tubes set up to investigate the effect of catechol oxidase (an enzyme present in apple cells) on catechol, its substrate, at room temperature.

tube	catechol	state of apple tissue	lead nitrate
1	√	fresh	√
2	√	boiled	√
3	√	boiled	×
4	√	fresh	×

(√ = present, × = absent)

Within a few minutes, the apple tissue would become dark brown in
   A tube 1 only.   **B** tubes 1 and 2.
   C tubes 3 and 4.   **D** tube 4 only.

Items **16**, **17** and **18** refer to the experiment shown in the diagram opposite, which was set up to measure a grasshopper's rate of respiration.
   After 30 minutes the coloured liquid in the experiment was returned to its original level by depressing the syringe plunger from point X to point Y.

16 The rise in level of coloured liquid indicates that the
   A grasshopper is giving out carbon dioxide.
   B sodium hydroxide is releasing carbon dioxide.
   C grasshopper is taking in oxygen.
   D sodium hydroxide is absorbing oxygen.

17 From this experiment it can be concluded that the grasshopper's rate of
   A oxygen consumption is 2.0 ml/hour.
   B oxygen consumption is 0.4 ml/hour.
   C carbon dioxide output is 0.2 ml/hour.
   D carbon dioxide output is 4.0 ml/hour.

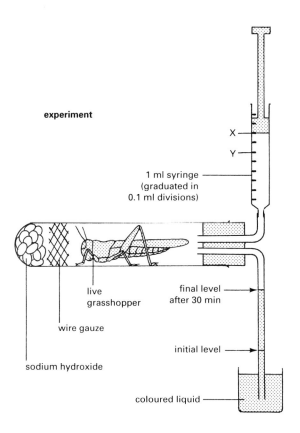
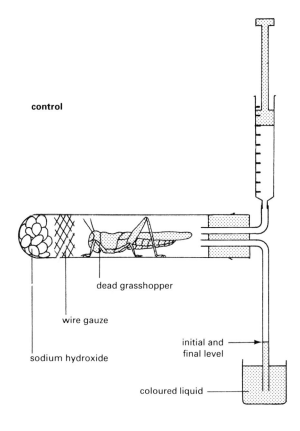

18 Which of the following procedures would *not* improve the reliability of the result?
  A replacing the dead grasshopper in the control tube with glass beads
  B repeating the experiment with the same grasshopper and calculating an average
  C pooling class results where each group used a different grasshopper
  D allowing a standard period of acclimatisation before taking readings

19 If an animal of mass 7 g consumes 3.5 cm³ of oxygen in 25 minutes, then its respiratory rate in cm³ oxygen used per gram of body tissue per minute is
  A 0.02   B 0.08   C 0.20   D 0.80

20 Plant cells respire 24 hours per day, but only photosynthesise when light is present. The following graph refers to a sample of *Elodea* (Canadian pond weed) over a period of 24 hours.

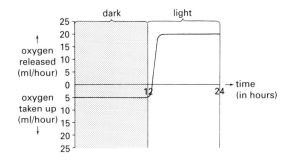

The most accurate estimate of the total volume of oxygen used by the plant for respiration during this 24-hour period is
  A 5 ml.   B 20 ml.   C 120 ml.   D 160 ml.

Test 5 Absorption of light by photosynthetic pigments

1. Of the total amount of solar energy falling on a leaf, the percentage used for photosynthesis is approximately
   A 5%.   B 25%.   C 50%.   D 75%.

2. In the diagram below of sunlight striking a green leaf, which arrow represents light being transmitted by the leaf?

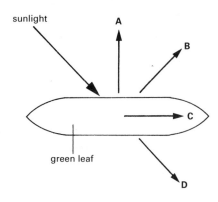

Items 3, 4, 5 and 6 refer to the experiment shown in the following diagram.

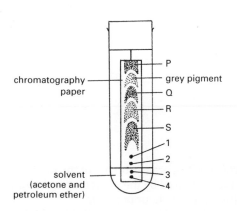

3. Two alternative positions at which the chlorophyll extract could have been spotted to give this separation are
   A 1 and 2.   B 1 and 3.
   C 2 and 3.   D 2 and 4.

4. Xanthophyll is at position
   A P.   B Q.   C R.   D S.

5. Chlorophyll a is at position
   A P.   B Q.   C R.   D S.

6. Carotene is at position
   A P.   B Q.   C R.   D S.

7. Which of the following diagrams best represents the absorption spectrum that results when a chlorophyll extract is placed in a beam of light?

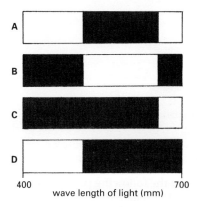

8. Which of the following graphs best represents the absorption spectrum of *both* green chlorophyll (Ch) *and* the yellow pigments (Y)?

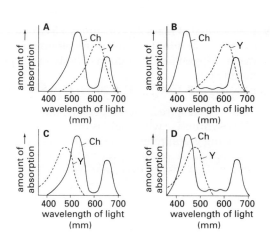

9 The two regions of the spectrum from which chlorophyll absorbs most light are
   A blue and yellow.   B green and blue.
   C red and green.   D red and blue.

10 The diagram below shows the result of an experiment in which a strand of alga was placed in water containing bacteria, and illuminated by a microspectrum of white light.

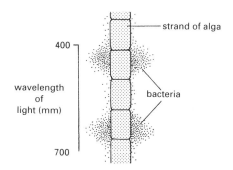

   Which of the following correctly explains the distribution of the bacteria?
   A The bacteria feed on the alga.
   B The alga receives carbon dioxide at these positions.
   C The bacteria receive oxygen at these positions.
   D The alga feeds on the bacteria.

11 Which of the following metal ions forms part of the structure of a chlorophyll molecule?
   A calcium   B copper
   C iron   D magnesium

Items 12, 13 and 14 refer to the following diagram which shows the structure of a chloroplast as revealed by an electron microscope.

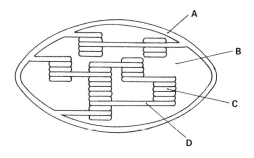

12 Which label line is pointing to the stroma?

13 Which labelled structure is the site of chlorophyll formation?

14 Which letter indicates the site of the carbon fixation stage of photosynthesis?

15 The diagram below shows a plant subjected in turn to four different sets of environmental conditions. Under which set of conditions would most photosynthesis occur?

A 55°C blue light   B 40°C green light   C 25°C red light   D 10°C daylight

Test 6 Chemistry of photosynthesis

1 The oxygen released by a green plant as a result of photolysis comes from
   A air.   B glucose.
   C water.   D carbon dioxide.

2 The light-dependent stage of photosynthesis results in the formation of two compounds needed for the carbon fixation stage. These are
   A adenosine triphosphate and reduced hydrogen acceptor.
   B reduced hydrogen acceptor and glycerate phosphate.
   C glycerate phosphate and ribulose bisphosphate.
   D ribulose bisphosphate and adenosine triphosphate.

13

3. Photophosphorylation is the name given to the process by which
   A. chemical energy is converted into light energy in grana.
   B. ADP and inorganic phosphate are formed by the breakdown of ATP.
   C. light energy is absorbed by photosynthetic pigments in grana.
   D. ATP is synthesised during the light-dependent stage of photosynthesis.

4. The first stable compounds resulting from the carbon fixation stage of photosynthesis are formed in the order
   A. glycerate phosphate → triose phosphate → hexose.
   B. hexose → glycerate phosphate → triose phosphate.
   C. glycerate phosphate → hexose → triose phosphate.
   D. hexose → triose phosphate → glycerate phosphate.

5. The graph below refers to an experiment involving a species of alga. The relative concentrations of GP (PGA) and RuBP (RuDP) present in the cells were monitored when the plants were in light, and then in darkness.

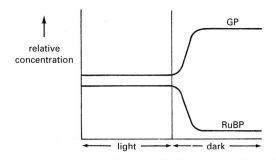

   Which of the following conclusions *cannot* be drawn from these results?
   A. In darkness the relative concentration of GP increases.
   B. During the experiment RuBP may be converted into GP.
   C. The relative concentration of RuBP decreases on removal of CO_2.
   D. In light a steady state exists between RuBP and GP.

Items 6, 7, 8, 9, 10 and 11 refer to the following diagram of the cyclic series of reactions that occurs during the carbon fixation stage of photosynthesis.

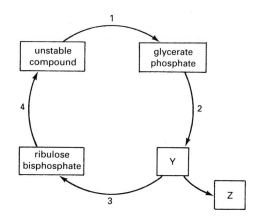

6. This metabolic pathway is also known as the
   A. tricarboxylic acid cycle.
   B. citric acid cycle.
   C. Krebs cycle.
   D. Calvin cycle.

7. Carbon dioxide is taken into the cycle at stage
   A. 1.   B. 2.   C. 3.   D. 4.

8. Hydrogen from reduced hydrogen acceptor is used at stage
   A. 1.   B. 2.   C. 3.   D. 4.

9. Energy from ATP is used to drive stages
   A. 1 and 2.   B. 2 and 3.
   C. 2 and 4.   D. 3 and 4.

10. The substance formed at position Y is
    A. 3-carbon sugar.   B. pyruvic acid.
    C. glucose-1-phosphate. D. citric acid.

11. If one molecule of substance Y is released per cycle, how many times must the cycle turn for one molecule of sucrose ($C_{12}H_{22}O_{11}$) to be built up at position Z?
    A. 2 times   B. 4 times
    C. 8 times   D. 12 times

12. Which of the following changes in concentration of chemicals would occur if an illuminated green plant cell's source of carbon dioxide were removed?

	ribulose bisphosphate	glycerate phosphate
A	increase	increase
B	decrease	decrease
C	increase	decrease
D	decrease	increase

14

13 The following graph presents the results from a photosynthesis experiment.

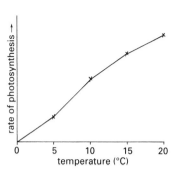

on the rate of photosynthesis at three different intensities of light.

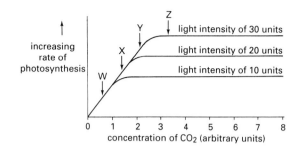

Which of the following pairs of environmental factors must have been kept constant in order to make this a fair test?
A light intensity and temperature
B temperature and carbon dioxide concentration
C water content and oxygen concentration
D light intensity and carbon dioxide concentration

Items **14** and **15** refer to the following graph which shows the effect of carbon dioxide concentration

14 At which of the following concentrations was carbon dioxide the limiting factor at all three light intensities?
A 0–1 units B 1–2 units
C 2–3 units D 3–4 units

15 Light intensity was the limiting factor at
A point W only.
B point Z only.
C points X and Y only.
D points W, X, Y and Z.

Test 7 DNA and its replication

1 DNA is *not* present in a
A nucleus.   B gene.
C membrane. D chromosome.

2 The sugar present in DNA is
A ribose.   B dextrose.
C ribulose.   D deoxyribose.

3 The structure of one nucleotide is shown below.

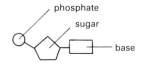

Which of the following diagrams shows two nucleotides correctly joined together?

4 Which of the following is a base pair normally present in DNA?
   A adenine and cytosine
   B guanine and adenine
   C thymine and guanine
   D thymine and adenine

Items **5** and **6** refer to the following table and list of possible answers.

cell types analysed	average mass of DNA/cell ($\times 10^{-12}$ g)
X	0.00
Y	3.35
kidney	6.70
lung	6.70

A sperm cell
B liver cell
C smooth muscle cell
D mature red blood cell

5 What is the correct identity of cell type X?

6 What is the correct identity of cell type Y?

7 The average mass of DNA present in an ovum of the species referred to in the above table would be
   A 3.35×10^{-6}
   B 6.70×10^{-6}
   C 3.35×10^{-12}
   D 6.70×10^{-12}

8 If a DNA molecule contains 10 000 base molecules, of which 18% are thymine, then the number of cytosine molecules present is
   A 1800.   B 3200.   C 6400.   D 8200.

9 A shorthand method of representing part of a single strand of DNA is shown below.

Which of the following shows the correct positions of the phosphate (P), sugar (S) and base (B) molecules?

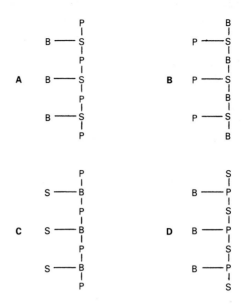

10 If a DNA molecule contains 4000 base molecules, and 1200 of these are adenine, then the percentage number of guanine bases present in the molecule is
    A 12.   B 20.   C 28.   D 30.

11 DNA molecules isolated from a rat cell and a human cell are found to differ in the order of their
    A bases only.
    B sugars only.
    C phosphates only.
    D bases, sugars and phosphates.

12 The following set of results shows an analysis of the DNA bases contained in the cells of a cow's thymus gland.

base composition			
X	guanine	Y	Z
28.2%	21.5%	21.2%	27.8%

Which of the following is a possible correct identification of the bases?

	X	Y	Z
A	cytosine	adenine	thymine
B	thymine	adenine	cytosine
C	adenine	cytosine	thymine
D	cytosine	thymine	adenine

13 During DNA replication the following events occur:
1 winding brings about formation of two double helices.
2 bases on free nucleotides bond with bases on the DNA strand.
3 hydrogen bonds break allowing DNA strands to unzip.
4 bonds form between adjacent nucleotide molecules.
   The correct order in which these events occur is
A 2, 3, 1, 4.   **B** 3, 2, 4, 1.
C 2, 3, 4, 1.   **D** 3, 4, 2, 1.

14 The following diagram shows a molecule of DNA prior to replication.

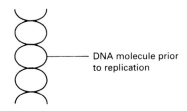

If ———— represents an original DNA strand, and – – – – represents a new DNA strand, which of the following daughter DNA molecules will result from replication of the DNA molecule shown above?

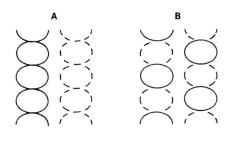

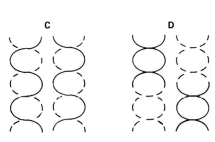

15 The diagram below shows the stages that occur in an actively dividing mammalian cell.
   If the drug aminopterin (which inhibits thymine formation) is added to a culture of actively dividing cells, at which stage in the cell cycle will most cells be present 16 hours after the addition of the drug?

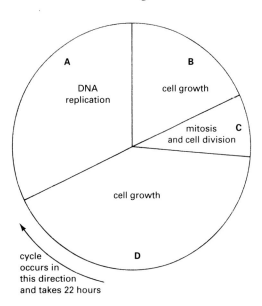

Test 8 RNA and protein synthesis

1 Which of the following is correct?

	present in DNA	present in RNA
A	uracil	thymine
B	deoxyribose	ribose
C	single strand	double strand
D	4 different nucleotides	5 different nucleotides

2 One of the nucleotides present in mRNA has the composition
  A adenine–ribose–phosphate.
  B uracil–deoxyribose–phosphate.
  C thymine–ribose–phosphate.
  D guanine–deoxyribose–phosphate.

3 Strand X in the diagram below shows a small part of a nucleic acid molecule.

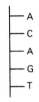

Which pair of the following strands are complementary to strand X?

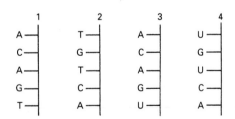

  A 1 and 3   B 2 and 4
  C 1 and 2   D 3 and 4

4 A mRNA template is
  A translated from protein.
  B transcribed into protein.
  C translated into DNA.
  D transcribed from DNA.

5 A free transfer RNA molecule can combine with
  A one specific amino acid only.
  B any available amino acid.
  C three different amino acids.
  D a chain of amino acids.

6 The number of bases present in one codon is
  A 1.   B 2.   C 3.   D 4.

Items 7, 8 and 9 refer to the possible answers numbered in the following list.
1 DNA molecules
2 tRNA molecules
3 mRNA molecules
4 amino acids
5 ribosomes

7 Which of these must be present in the largest number for successful synthesis of a large protein molecule to occur?
  A 1   B 3   C 4   D 5

8 Which of these controls the order in which amino acids are added to a growing protein chain?
  A 2   B 3   C 4   D 5

9 On which of these are anti-codons present?
  A 2   B 3   C 4   D 5

10 If each amino acid weighs 100 mass units, what is the weight (in mass units) of the protein molecule synthesised from a mRNA molecule which is 600 bases long?
  A 2000   B 6000   C 20 000   D 60 000

11 The table below shows three different mRNA molecules (each containing a base sequence) and the three different protein molecules synthesised from them.

mRNA	repeating sequence	protein
AGAGAGAGAGAGAGAG – – – – –	AG	X
CAUCAUCAUCAUCAUCAU – – – –	CAU	Y
AAUGAAUGAAUGAAUGAAUGAAUG –	AAUG	Z

Which of the following shows the correct number of different types of amino acid in each protein molecule?

	X	Y	Z
A	2	1	4
B	1	3	2
C	2	1	3
D	3	1	4

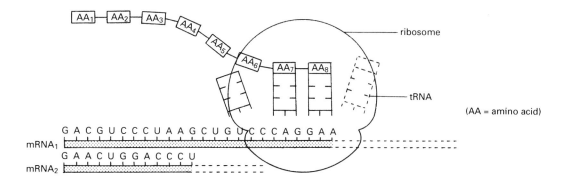

Items **12**, **13** and **14** refer to the above diagram which shows the synthesis of part of a protein molecule.

12 Which of the following is the first part of the protein molecule that would be translated from mRNA$_2$?

start of protein ↓

A AA$_4$–AA$_2$–AA$_7$–AA$_6$ – – – – – – –
B AA$_6$–AA$_7$–AA$_2$–AA$_4$ – – – – – – –
C AA$_3$–AA$_1$–AA$_5$–AA$_8$ – – – – – – –
D AA$_8$–AA$_5$–AA$_1$–AA$_3$ – – – – – – –

13 The DNA strand from which mRNA$_2$ was synthesised is
A GAACTGGACCCT.
B CTTGACCTGGGA.
C GAACUGGACCCU.
D CUUGACCUGGGA.

14 The following diagram shows a small part of a different protein that was also synthesised on this ribosome.

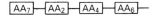

What sequence of bases in DNA coded for this sequence of amino acids?
A CAGGUCAAGUCC
B CAGCTCAAGTCC
C GUCCAGUUCAGG
D GTCCAGTTCAGG

15 Ribosomes are found to occur
A freely in the cytoplasm and attached to the nuclear membrane.
B attached to the endoplasmic reticulum and to the nuclear membrane.
C freely in the cytoplasm and attached to the endoplasmic reticulum.
D attached to the Golgi apparatus and to the endoplasmic reticulum.

16 Which of the following is composed of a system of flattened sacs and tubules?
A nuclear membrane
B mitochondrion
C endoplasmic reticulum
D ribosome

17 In which of the following would the Golgi apparatus be most highly developed?
A salivary gland cells
B red blood corpuscles
C skeletal muscle fibres
D kidney tubule cells

18 The relative number of ribosomes would be greatest in
A sieve tubes in a ripe fruit.
B xylem vessels in a woody stem.
C epidermal cells in a mature root.
D leaf cells in a growing bud.

19 Which of the following statements about the Golgi apparatus is *incorrect*?
A It is a group of flattened fluid-filled sacs.
B It is formed from vesicles pinched off from the endoplasmic reticulum.
C It contains newly synthesised protein.
D It is attached to the nuclear membrane.

20 Prior to mucus (a type of protein) leaving a goblet cell and playing its role in the trachea, the following events occur.
1 fusion of vesicle with plasma membrane
2 addition of carbohydrate component to protein
3 secretion of processed glycoprotein by cell
4 separation of vesicle from Golgi apparatus
The sequence in which these events occur is
A 4, 2, 3, 1. B 2, 4, 1, 3.
C 4, 2, 1, 3. D 2, 4, 3, 1.

Test 9 Variety of proteins

1. Which of the following chemical elements is present in proteins but *not* in fats?
   A carbon
   B oxygen
   C nitrogen
   D hydrogen

2. Which of the following chemical elements is often a constituent of protein?
   A calcium
   B sodium
   C potassium
   D sulphur

3. The number of different types of amino acid commonly found to make up proteins is approximately
   A 20.   B 30.   C 200.   D 1000.

Items 4 and 5 refer to the following possible answers.
A fibrous protein   B hydrolysed protein
C conjugated protein D globular protein

4. Which of these consists of polypeptide chains arranged in long parallel strands?

5. Which of these is illustrated by the following diagram?

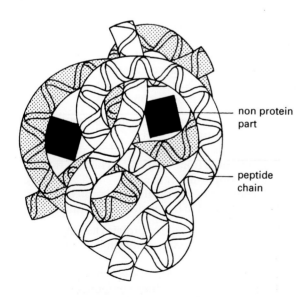

6. Which of the following is an example of a conjugated protein?
   A insulin
   B haemoglobin
   C pepsin
   D glucagon

7. Three proteins found in the human body are
   1 thyroxine,
   2 cytochrome oxidase,
   3 immunoglobulin (antibody).
   Which of the following correctly identifies their functions?

	function		
	regulates growth and metabolism	defends the body against microbes	speeds up rate of a biochemical process
A	1	2	3
B	2	3	1
C	1	3	2
D	3	1	2

8. The following table gives the mass per 100 g of protein of five different amino acids found in four proteins.

	protein	mass of amino acid (g/100 g protein)				
		glycine	alanine	leucine	valine	phenyl-alanine
A	insulin	4	5	13	8	8
B	haemoglobin	6	7	15	9	8
C	keratin	7	4	11	5	4
D	albumin	3	7	9	7	8

Which protein is represented by the following pie chart?

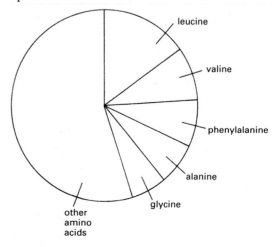

9 The following diagram shows the sequence of amino acids present in one molecule of insulin.

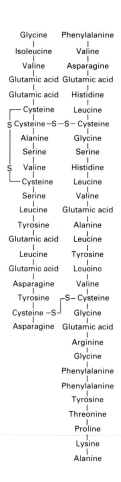

In this protein the ratio of leucine : glycine : tyrosine : histidine is

A 6 : 4 : 3 : 2
B 6 : 4 : 4 : 1
C 3 : 2 : 2 : 1
D 3 : 2 : 1 : 1

10 In the following diagram of enzyme action, which structure represents the enzyme?

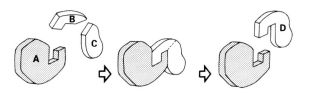

Test 10 Viruses

1 Which labelled structure in the following diagram of a virus is nucleic acid?

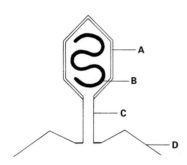

2 It is *correct* to say that viruses
   A reproduce both inside and outside living cells.
   B exhibit living and non-living characteristics.
   C are always transmitted from animal to animal by blood contact.
   D are always transmitted from plant to plant by aphids.

3 The diameter of a certain virus is 20 nanometres (nm). If 1 micrometre (μm) = 1000 nanometres, then the diameter of this virus, expressed in metres (m), is
   A 2×10^{-8}.   B 2×10^{8}.
   C 2×10^{-9}.   D 2×10^{9}.

4 In the following comparison, which pair of statements is *not* correct?

	virus	unicellular alga
A	contains one type of nucleic acid	contains two types of nucleic acid
B	depends on host cell for synthesis of nucleic acid and protein coats	is able to synthesise all of its own nucleic acid and protein molecules
C	cannot form the first link in a food chain	can form the first link in a food chain
D	always arises directly from another virus	never arises directly from a pre-existing cell

Items **5** and **6** refer to the following diagram which shows some of the stages that occur during viral invasion of a bacterium.

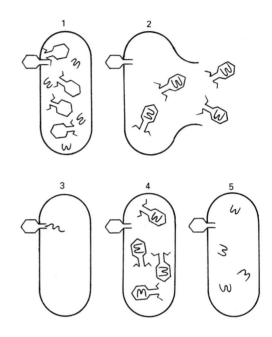

5 The sequence in which these events would occur is
   A 3,1,5,4,2.   B 2,4,1,5,3.
   C 3,5,1,4,2.   D 2,5,1,3,4.

6 Which numbered stage illustrates viral nucleic acid molecules about to enter newly formed protein coats?
   A 1   B 2   C 4   D 5

virus	nucleic acid	protein coat
coliphage fd	single-stranded DNA	naked and helical
pox virus	double-stranded DNA	enveloped and helical
herpes virus	double-stranded DNA	enveloped and polyhedral
adenovirus	double-stranded DNA	naked and polyhedral
tobacco mosaic	single-stranded RNA	naked and helical
picornavirus	single-stranded RNA	naked and polyhedral
myxovirus	single-stranded RNA	enveloped and helical
reovirus	double-stranded RNA	naked and polyhedral

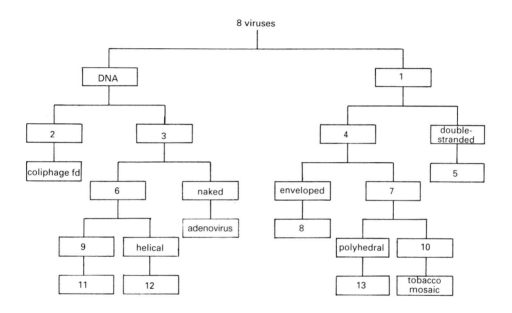

Items **7**, **8**, **9** and **10** refer to the branched key above which was constructed using the information in the table at the top of the page.

7 The characteristic 'double-stranded' should be at position
A 1. **B** 2. **C** 3. **D** 4.

8 The characteristic 'helical' should be at position
A 6. **B** 7. **C** 9. **D** 10.

9 Pox virus would be correctly classified at position
A 5. **B** 8. **C** 11. **D** 12.

10 The virus at position 13 should be
A herpes. **B** myxovirus.
C picornavirus. **D** reovirus.

Test 11 Cellular response in defence

1. Which of the following statements is *correct*?
   A Interferon molecules act directly on invading virus particles.
   B Many plants produce interferon in response to viral infection.
   C Each interferon molecule produced is specific to a particular virus.
   D Interferon triggers cells to produce antiviral proteins.

2. Ideal conditions for the growth of pathogens in the human body are provided by a supply of
   A food, carbon dioxide and warmth.
   B moisture, food and carbon dioxide.
   C warmth, moisture and food.
   D carbon dioxide, moisture and warmth.

3. In humans, first line defences against microbes are provided by *both*
   A phagocytes and stomach acid.
   B skin and lysozyme in tears.
   C antibodies and clotting of blood.
   D white blood cells and vomiting.

4. In the table below, which method of defence against viral invasion has its type of immune response *incorrectly* classified?

	method of defence	type of immune response	
		non-specific	specific
A	phagocytosis	✓	
B	immunisation	✓	
C	antibody formation		✓
D	interferon	✓	

5. The human body is defended against viral invasion by
   A B-lymphocytes functioning as phagocytes which engulf viruses.
   B neutrophils producing free antibodies which act against viruses.
   C monocytes functioning as phagocytes which produce free antibodies.
   D T-lymphocytes producing cell-bound antibodies which act against viruses.

6. The following diagram shows four of the stages which occur during the process of phagocytosis.

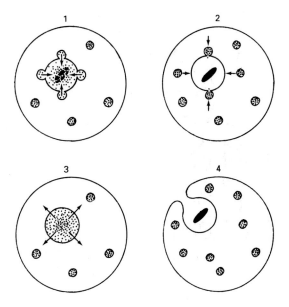

Which set in the table below correctly matches each of these numbered stages with its description?

description of stage	set			
	A	B	C	D
products of digestion pass into cytoplasm of phagocyte	3	4	2	1
some lysosomes move towards and fuse with vacuole	2	1	3	4
phagocyte forms vacuole around bacterium	4	3	1	2
digestive enzymes break bacterium down	1	2	4	3

7 When a virus with antigens on its surface invades an organism and multiples, the following events occur.
   1 Many copies of antibodies are synthesised by white blood cells.
   2 Virus particles become attached by antigens to white blood cells.
   3 Virus antigens combine with antibodies.
   4 White blood cells become triggered and multiply.
   The order in which these occur is
   A 3, 2, 1, 4.   **B** 2, 4, 1, 3.
   C 4, 1, 3, 2.   **D** 2, 1, 4, 3.

8 Which of the diagrams below shows the correct structure of an antibody molecule? (R = receptor site, N = non-receptor site.)

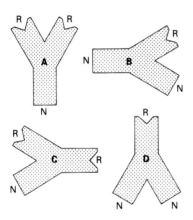

9 A person acquires long-lasting natural immunity to a particular antigen when she or he
   A receives a small amount of vaccine by injection.
   B is given antibodies produced by another mammal.
   C responds to the antigen by producing antibodies.
   D receives antibodies as a baby during suckling.

10 The following table summarises the physiological differences between the four types of human blood group.

blood group	antigens on red corpuscles	antibodies present in plasma
A	A	Anti-B
B	B	Anti-A
AB	A and B	neither
O	neither	Anti-A and Anti-B

When red blood corpuscles carrying one or both antigens are donated to a recipient with the corresponding antibodies in her or his plasma, the transfusion is unsuccessful because agglutination (clumping) of red corpuscles occurs.

The table below presents the feasibility of blood transfusions between people of the same and different blood groups. (+ = successful transfusion, − = unsuccessful transfusion.)

donor's blood group	recipient's blood group			
	A	B	AB	O
A	+	1	+	−
B	−	+	+	2
AB	3	−	+	−
O	+	+	4	+

Which of the numbered transfusions in the table would be successful?
A 1   **B** 2   **C** 3   **D** 4

11 Choose the correct identities of the three numbered blanks in the following sentence, from the answers below.
   Immuno-suppressor drugs prevent the ___1___ of a transplanted tissue from making ___2___, which would attack ___3___ present in the transplanted tissue.

	1	2	3
A	recipient	antibodies	antigens
B	donor	antibodies	antigens
C	recipient	antigens	antibodies
D	donor	antigens	antibodies

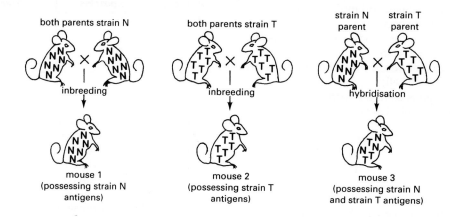

12 The diagram above shows the crosses which resulted in the production of three mice (1, 2 and 3).
   Which of the following skin grafts would be *least* likely to be rejected?

	donor mouse	recipient mouse
A	3	1
B	2	1
C	1	3
D	3	2

13 Phytoalexins are produced by many
   A plants as a defence against fungal attack.
   B animals as a defence against fungal attack.
   C plants as a defence against viral attack.
   D animals as a defence against viral attack.

14 Which of the following does *not* correctly describe a defence mechanism adopted by plants against invasion by other organisms?
   A Galls isolate parasites from the rest of the plant.
   B Ethylene gas speeds up abscission of leaves damaged by viruses.
   C Extra lignin deposits resist penetration by fungal threads.
   D Tannins block off sieve tubes preventing spread of bacteria.

15 The following table summarises a series of experiments involving the sawfly which causes 'bean galls' on the leaves of willow trees.

details of experiment	result
fertile female allowed to puncture leaf and lay eggs which hatch	+
fertilised eggs extracted from female and implanted into leaf tissues	−
fertile female allowed to pierce leaf and lay eggs which are then killed using a needle	+
adult males left in contact with gall-free willow leaves	−
fertile female allowed to puncture leaf but prevented from laying eggs	+
hungry larvae removed from galls and placed on gall-free leaf	−
sterilised female allowed to pierce leaf and lay sterile eggs which fail to hatch	+

(+ = gall produced, − = no gall produced)

Which of the following hypotheses is supported by this series of experiments?

In this particular relationship between gall-causer and gall-maker, gall formation
   A is stimulated by a chemical released by the digestive glands of feeding larvae.
   B depends on a chemical present on the surface of newly laid eggs.
   C is induced by a chemical injected by the female at the time of egg-laying.
   D only occurs if eggs release a chemical on hatching into larvae.

Test 12 Meiosis

1 The following diagram shows the formation of an animal zygote.

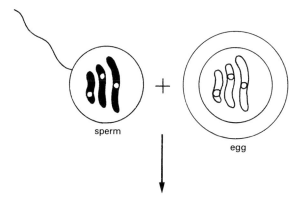

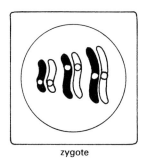

This zygote contains
A 6 chromatids and is haploid.
B 6 chromatids and is diploid.
C 3 pairs of chromosomes and is haploid.
D 3 pairs of chromosomes and is diploid.

2 Meiosis
A involves only one cell division.
B produces identical daughter cells.
C increases variation in a population.
D produces new body cells during repair.

3 Replication of the deoxyribonucleic acid (DNA) necessary for meiosis occurs
A while the chromosomes are arranged at the equator.
B before the chromatids appear.
C between the first and second divisions.
D after the homologous chromosomes become separated.

4 Which of the products shown in the following table results from meiosis in a sperm mother cell (ploidy number = 2n)?

	number of sperm formed	ploidy number of each sperm
A	2	n
B	2	2n
C	4	n
D	4	2n

5 The following diagram shows the nuclear contents of a cell.

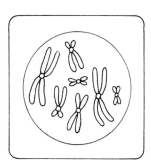

Which of the following descriptions refers accurately to this cell?

	number of chromosomes	number of pairs of homologous chromosomes
A	7	3
B	7	7
C	14	3
D	14	7

27

6 The following diagram shows a sperm mother cell from a fruit fly. The paternal chromosomes are shaded and the maternal chromosomes are unshaded.

The chance of a sperm receiving all four maternal chromosomes is
A 1 in 2. B 1 in 4.
C 1 in 8. D 1 in 16.

7 Which of the following correctly shows a pair of homologous chromosomes at the start of meiosis?

A
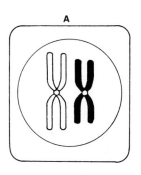

B
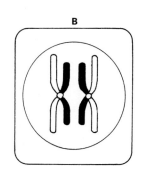

C
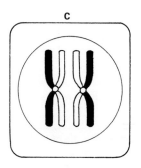

D

8 The following diagram shows a cell undergoing meiosis.

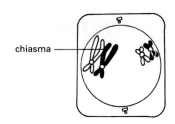

Which of the diagrams below shows the next stage in the process?

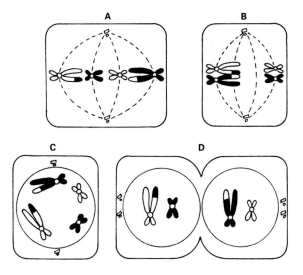

9 In the following diagram of a cell undergoing meiosis, assume that crossing-over occurs only at the chiasma indicated.

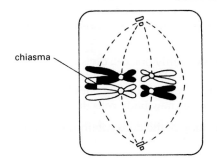

Which of the following gametes will *not* be formed from this cell?

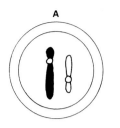
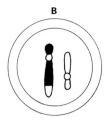
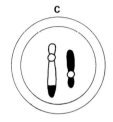
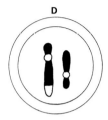

10 In which of the following do *both* structures contain cells which divide by meiosis?
A stigma and stamen
B ovary and anther
C stamen and style
D anther and stigma

11 Which of the following could be used to prepare a microscope slide of cells undergoing meiosis?
A locust testis
B pea root tip
C human cheek epithelium
D fruit fly salivary gland

12 The following diagram represents the human life cycle.

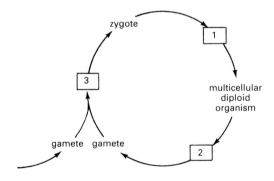

Which of the following combinations of terms correctly matches the numbered boxes?

	1	2	3
A	mitosis	fertilisation	meiosis
B	meiosis	mitosis	fertilisation
C	fertilisation	meiosis	mitosis
D	mitosis	meiosis	fertilisation

13 Four different steps that occur during meiosis are given in the following list.
1 complete separation of chromatids
2 pairing of homologous chromosomes
3 lining up of paired chromosomes on equator
4 crossing-over between chromatids
These steps would occur in the order
A 2, 3, 4, 1.   B 3, 2, 4, 1.
C 2, 4, 3, 1.   D 3, 1, 2, 4.

14 Each egg mother cell in a dandelion plant contains 24 chromosomes. The number of chromosomes present in a root tip cell would be
A 6.   B 12.   C 24.   D 48.

15 Which of the following statements is true of mitosis but *not* of meiosis?
A The chromosome number is unaltered by the process.
B The cells produced by the process differ from each other.
C Variation within a population is increased by the process.
D The cells produced by the process are all haploid.

Test 13 Monohybrid cross

1 In humans, the condition hyperdactyly (the possession of twelve fingers) is determined by a dominant allele (H) and the normal condition by the recessive allele (h).
  The following diagram shows a family tree in which some members of the family are hyperdactylous.

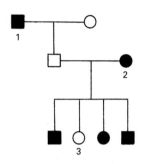

■ = hyperdactylous male
□ = normal male
● = hyperdactylous female
○ = normal female

The genotypes of persons 1, 2 and 3 in this family tree are

	1	2	3
A	Hh	Hh	hh
B	HH	HH	hh
C	Hh	HH	Hh
D	HH	Hh	hh

2 In *Drosophila*, the allele for normal grey body colour (E) is dominant to the allele for ebony body colour (e). The following table summarises the results of several crosses.

cross	result
strain 1 × ee	all wild type
strain 2 × ee	1 wild type : 1 ebony
strain 3 × ee	all ebony
strain 4 × Ee	3 wild type : 1 ebony

Which strains *both* have the genotype Ee?
A 1 and 3   B 1 and 4
C 2 and 3   D 2 and 4

3 In maize plants, two alleles of the gene for seed colour exist. Purple (P) is dominant to yellow (p).
  A backcross (testcross) was carried out to determine the genotype of a certain purple plant. Which of the following is correct?

	phenotypic ratio of offspring resulting from backcross	genotype of purple parent
A	1 purple : 1 yellow	heterozygous
B	3 purple : 1 yellow	homozygous
C	1 purple : 1 yellow	homozygous
D	all purple	heterozygous

4 In mice, Y is the dominant allele for yellow fur and y is the recessive allele for grey fur. Since Y is lethal when homozygous, the result of cross Yy × Yy will be
A 3 yellow : 1 grey.   B 2 yellow : 1 grey.
C 1 yellow : 1 grey.   D 1 yellow : 2 grey.

5 A certain type of anaemia exists in two forms, major (severe) and minor (mild). The following table relates the genotypes of both types of sufferer to that of a normal person.

person	genotype
non sufferer	NN
minor sufferer	NA
major sufferer	AA

If NA marries NA, the chance of each of their children being mildly infected is
A 1 in 1.   B 1 in 2.
C 1 in 3.   D 1 in 4.

Items **6** and **7** refer to the following information.
  In shorthorn cattle, coat colour is controlled by a gene which shows incomplete dominance. RR = red, RW = roan and WW = white.
  The following diagram shows the outcome of a cross between two members of this type of cattle.

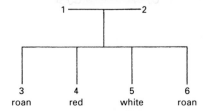

30

6 The phenotypes of parents 1 and 2 are

	1	2
A	roan	white
B	white	red
C	red	roan
D	roan	roan

7 Which two individuals when crossed will produce offspring of only one phenotype?
A 3 and 4
B 3 and 6
C 4 and 5
D 5 and 6

8 Allele O is recessive to both A and B, the equally dominant alleles of the gene determining human blood group. In which of the following paternity suits is the alleged father clearly *not* the child's father?

paternity suit	mother's genotype	alleged father's genotype	child's genotype
A	BO	AA	AB
B	AB	BO	BB
C	AO	OO	BO
D	OO	AB	AO

Items **9** and **10** refer to the following family tree.

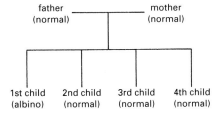

9 If A = normal allele and a = albino allele, the genotypes of these parents are

	father	mother
A	Aa	Aa
B	AA	AA
C	AA	Aa
D	Aa	AA

10 The chance of this couple's fifth child being an albino is
A 1 in 1.
B 1 in 2.
C 1 in 3.
D 1 in 4.

11 In snapdragon plants, the alleles for red and ivory flower colour show incomplete dominance.
 When a homozygous red-flowered plant is crossed with a homozygous ivory-flowered plant, all the members of the F_1 generation are found to bear pink flowers.
 Which of the following would be the outcome of crossing a red-flowered plant with a pink one?
A 1 red : 2 pink : 1 ivory
B 3 red : 1 ivory
C 1 red : 1 pink
D all red

12 In mice, black coat colour (allele B) is dominant to brown coat colour (allele b). The offspring of a cross between a black mouse (BB) and a brown mouse were allowed to interbreed. What percentage of the progeny would have black coats?
A 25%   B 50%   C 75%   D 100%

13 In fruit flies, long wing is dominant to vestigial wing. When heterozygous long-winged flies were crossed with vestigial-winged flies, 192 offspring were produced. Of these, 101 had long wings and 91 had vestigial wings.
 If an exact Mendelian ratio had been obtained, then the number of each phenotype would have been

	long-winged	vestigial-winged
A	64	128
B	96	96
C	128	64
D	192	0

Items **14** and **15** refer to the following information and the list of possible answers after the table (overleaf).
 In rabbits, homozygous combinations of three alleles of the gene for coat colour produce the phenotypes:
  CC = brown
  $C^{ch}C^{ch}$ = chinchilla
  C^HC^H = Himalayan
C is completely dominant to C^{ch}, which is completely dominant to C^H.

The results of two crosses are given in the following table.

cross	phenotypes of parents	phenotypes of progeny		
		brown	chinchilla	Himalayan
1	brown × Himalayan	8	9	0
2	brown × chinchilla	18	7	9

A $CC^{ch} \times C^H C^H$
B $CC^{ch} \times C^{ch} C^H$
C $CC^H \times C^H C^H$
D $CC^H \times C^{ch} C^H$

14 Which of the above answers gives the genotypes of both the cross 1 parents?

15 Which of the above answers gives the genotypes of both the cross 2 parents?

Test 14 Dihybrid cross

Items **1** and **2** refer to the following information and list of possible answers.

In a certain plant, yellow fruit colour (Y) is dominant to green (y) and round shape (R) is dominant to oval (r). The two genes involved are located on different chromosomes.

A 9:3:3:1 ratio of phenotypes only
B 9:3:3:1 ratio of genotypes only
C 1:1:1:1 ratio of phenotypes only
D 1:1:1:1 ratio of phenotypes and genotypes

1 Which of the above will result when plant YyRr is self-pollinated?

2 Which of the above will result when plant YyRr is backcrossed (testcrossed) with the double recessive parent?

Items **3** and **4** refer to the following information.

In *Drosophila*, dumpy wing (d) is recessive to normal wing (D) and ebony body (e) is recessive to normal body colour (E). The two genes involved are not on the same chromosome.

A true-breeding normal-winged, ebony-bodied fly is crossed with a true-breeding dumpy-winged, normal-bodied fly.

3 The genotype of the F_1 generation will be
A Dd/Ee.   B DD/Ee.
C DD/EE.   D Dd/EE.

4 As a result of interbreeding amongst the members of the F_1 generation, dumpy-winged, normal-bodied flies will be present in the F_2 generation in the proportion
A 1 in 16.   B 3 in 16.
C 6 in 16.   D 9 in 16.

5 In mice, black coat (allele B) is dominant to white coat (b) and straight whiskers (allele S) is dominant to curved whiskers (s).

When true-breeding mice with black coats and straight whiskers were crossed with white mice possessing curved whiskers, the offspring were all black with straight whiskers.

If these F_1 mice were crossed with white mice possessing curved whiskers, the expected proportion of offspring with black coats and curved whiskers in the next generation would be
A 1 in 16.   B 3 in 16.
C 4 in 16.   D 9 in 16.

6 When a true-breeding plant bearing disc-shaped yellow fruit was crossed with a plant bearing sphere-shaped green fruit, the F_1 generation were all found to bear disc-shaped yellow fruit.

This F_1 generation was self-pollinated and produced 800 offspring. Predict which of the following proportions of phenotypes was found in this F_2 generation.

	phenotype			
	disc-shaped yellow	disc-shaped green	sphere-shaped yellow	sphere-shaped green
A	199	201	202	198
B	49	153	152	446
C	800	0	0	0
D	453	148	147	52

7 In a certain species of sweet pea plant, flowers are either purple or white. Colour is determined by two unlinked genes. The alleles of the first gene are X and x; those of the second gene are Y and y.

In order to bear purple flowers, a plant must possess at least one X and one Y allele. Those genotypes which fail to do so, result in the formation of white flowers.

If two purple-flowered plants of genotype XxYy are crossed then the expected phenotypic ratio of offspring would be
A 12 purple:4 white.   B 9 purple:7 white.
C 10 purple:6 white.   D 8 purple:8 white.

8 In snapdragon plants, the broad leaf is completely dominant to narrow leaf, whereas red flower colour is incompletely dominant to ivory. (The genes for leaf width and flower colour are not linked.)

If a plant which is heterozygous for both genes is crossed with a true-breeding broad-leaved red-flowered plant, then the expected proportion of broad-leaved plants with pink flowers amongst the offspring would be
A 1 in 4.   B 2 in 4.
C 3 in 4.   D 4 in 4.

9 The diagram below shows a pair of homologous chromosomes during meiosis.

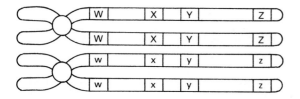

Most crossing over will occur between genes
A W and X.   B X and Y.
C W and Z.   D Y and Z.

10 The following diagram shows a homologous (bivalent) pair of chromosomes during meiosis.

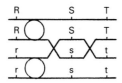

Which of the following correctly represents the final products of the second meiotic division?

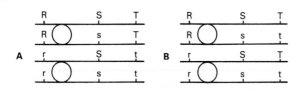

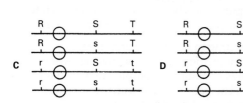

Items **11** to **13** refer to the following information.
In humans, the gene for red blood corpuscle shape (alleles elliptical E and normal e) is linked to the gene for Rhesus blood (alleles Rhesus positive R and Rhesus negative r).

11 A person with alleles E and R on one chromosome and e and r on its homologous partner will definitely produce gametes with the genotypes
A Ee and Rr.   B Ee and er.
C ER and Rr.   D ER and er.

12 If crossing-over occurs between these two genes, then the two additional types of gametes that could result are
A RE and re.   B EE and rr.
C Er and eR.   D ee and RR.

13 Person EeRr (referred to in item **11**) marries person eerr. If linkage between the two genes is *complete*, what is the chance of this couple producing a child with the genotype Eerr?
A 0 in 4   B 1 in 4   C 2 in 4   D 2 in 4

14 The following crosses refer to experiments using the fruit fly, *Drosophila*.
True-breeding, red-eyed flies with plain thoraxes were crossed with pink-eyed flies with striped thoraxes. The F_1 flies were then testcrossed against the double recessive as follows:

$$\begin{array}{c} \text{red eye} \\ \text{plain thorax} \end{array} \times \begin{array}{c} \text{pink eye} \\ \text{striped thorax} \end{array}$$

The following F_2 generation resulted from the cross.

80	16	12	92
red eye plain thorax	red eye striped thorax	pink eye plain thorax	pink eye striped thorax

What percentage number of recombinants resulted from the testcross?
A 12   B 14   C 16   D 28

15 In a certain species of animal, genes T, U, V and W occur on the same chromosomes. The following table gives their cross-over values (COVs).

linked gene pair	COV
T and U	25
T and V	5
V and U	30
U and W	10
V and W	20

Which of the following represents the correct order of the genes on the chromosome?
A V, T, W, U.   B T, W, U, V.
C T, V, W, U.   D V, W, T, U.

Test 15 Sex linkage

1. Which of the following male animals is *not* heterogametic?

		chromosome complement
A	fruit fly	2n = 6 + XY
B	fowl	2n = 14 + XX
C	grasshopper	2n = 16 + XO
D	human	2n = 44 + XY

Items **2** and **3** refer to the chromosome complements of male and female fruit flies, shown in the following diagram.

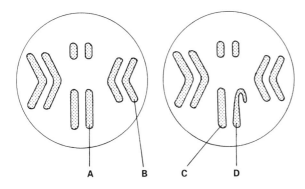

2. By which letter is an autosome labelled?

3. By which letter is a Y chromosome labelled?

4. A sex-linked allele *never* passes from a
   A man to his sons.
   B woman to her daughters.
   C man to his grandsons.
   D woman to her granddaughters.

Items **5** and **6** refer to eye colour in the fruit fly. In this sex-linked trait, the allele for red eye is dominant to that for white eye.

5. If a red-eyed male is crossed with a white-eyed female, their offspring will occur in the ratio
   A 1 red-eyed female : 1 red-eyed male.
   B 1 white-eyed female : 1 red-eyed male.
   C 1 red-eyed female : 1 white-eyed male.
   D 1 red-eyed female : 1 white-eyed female : 1 red-eyed male : 1 white-eyed male.

6. If a heterozygous red-eyed female is crossed with a white-eyed male, what percentage of the female offspring will be white-eyed?
   A 0%   B 25%   C 50%   D 100%

7. Haemophilia is a condition in which blood fails to clot, or clots only very slowly. Studies of this human sex-linked trait show that
   A every X chromosome carries the dominant allele.
   B a Y chromosome never carries the dominant allele.
   C both X and Y chromosomes can bear the recessive allele.
   D neither X nor Y chromosomes can bear the recessive allele.

8. Which of the following offspring could be produced by a normal homozygous female and a haemophiliac male?
   A normal males and normal females
   B haemophiliac males and normal females
   C normal males and carrier females
   D haemophiliac males and carrier females

9. A human female will definitely be a haemophiliac if
   A both of her parents are also haemophiliacs.
   B her mother is a carrier and her father is a haemophiliac.
   C her mother carries the allele for haemophilia on both X chromosomes.
   D her father is a haemophiliac and her mother is normal.

Items **10**, **11** and **12** refer to the following information and family tree.

In humans, type of tooth enamel is a sex-linked trait. Brown tooth enamel (e) is a recessive to normal (E).

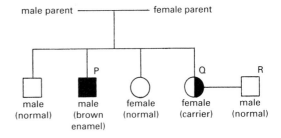

10 Person P's genotype is
   A X^eX^e.   B X^eY^e.   C XY^e.   D X^eY.

11 The genotypes of the two parents are
   A X^EY and X^EX^e.   B X^eY and X^EX^E.
   C X^eY and X^EX^e.   D X^EY and X^eX^e.

12 The probability of a child produced by persons Q and R being a boy with brown tooth enamel is
   A 1 in 1.   B 1 in 2.
   C 1 in 3.   D 1 in 4.

13 In cattle, the male is heterogametic. When a normal male is crossed with a female heterozygous for a sex-linked lethal gene, the sex ratio of their living offspring will be

	female : male
A	3 : 1
B	2 : 1
C	1 : 1
D	1 : 2

14 Red-green colour-blindness is a sex-linked trait in humans. X^C = normal allele and X^c = colour-blind allele in the following cross.

$$X^CX^c \times X^cY$$
$$\downarrow$$
$$X^cY$$
(Pat)

Which of the groups shown in the following consists of Pat's grandparents?

	maternal grand-mother	maternal grand-father	paternal grand-mother	paternal grand-father
A	X^CX^C	X^CY	X^CX^c	X^cY
B	X^CX^c	X^CY	X^CX^C	X^cY
C	X^CX^c	X^cY	X^CX^c	X^CY
D	X^CX^c	X^cY	X^CX^c	X^cY

15 In the magpie moth, wing colour is controlled by a sex-linked gene (where normal wing colour is dominant to pale colour).
   In poultry (Light Sussex variety), plumage colour is controlled by a sex-linked gene (where white is dominant to red).
   In the parental generation (P) of each of the following crosses, the homogametic sex is homozygous for the colour gene.

magpie moth	
P	pale male × normal female
	↓
F_1	1 normal male : 1 pale female

poultry	
P	red male × white female
	↓
F_1	1 white male : 1 red female

From these results it can be concluded that the heterogametic sex is
A male in magpie moth and female in poultry.
B female in magpie moth and male in poultry.
C male in both magpie moth and poultry.
D female in both magpie moth and poultry.

Test 16 Mutation

1. A mutation is a
   A sudden temporary change in an organism's genetic material.
   B change in phenotype followed by a change in genotype.
   C change in hereditary material directed by a changing environment.
   D change in genotype which may result in a new expression of a characteristic.

2. Which of the following statements is *not* correct?
   A Mutations provide variation upon which natural selection can act.
   B The vast majority of mutations produce alleles which are dominant.
   C Mutations arise spontaneously, infrequently and at random.
   D Mutation rate can be increased by artificial means.

3. A comparison of the karyotypes of a normal human male and a male sufferer of Down's syndrome shows the latter to possess
   A one extra chromosome.
   B two Y chromosomes.
   C one extra pair of chromosomes.
   D twice the total number of chromosomes.

4. The difference in karyotypes referred to in item 3 is the result of
   A non-disjunction.   B polyploidy.
   C sex linkage.   D chiasmata.

5. Polyploid wheat does *not* normally show an increase in
   A size.
   B vigour.
   C resistance to disease.
   D length of life cycle.

Items **6** and **7** refer to the following information.
The formation of the fertile polyploid plant *Spartina townsendii* from *Spartina maritima* (2n = 56) and *Spartina alterniflora* (2n = 70) is thought to have involved
1 vegetative growth of a sterile hybrid,
2 spindle failure in a meristematic cell,
3 fertilisation between gametes of two different species.

6. If this theory is true, then these intermediate steps must have occurred in the order
   A 3, 2, 1.   B 2, 1, 3.
   C 3, 1, 2.   D 2, 3, 1.

7. The chromosome complement of *Spartina townsendii* is
   A 2n = 63.   B 2n = 91.
   C 2n = 98.   D 2n = 126.

8. The following table shows the chromosome numbers of several species of *Potentilla*. Species 1 (*Potentilla rupestris*) has the basic chromosome number. All of the other species are polyploids.

species	haploid chromosome number	diploid chromosome number
1 Potentilla rupestris	n = 7	2n = 14
2 Potentilla anserina	n = 14	2n = 28
3 Potentilla crantzii	n = 21	2n = 42
4 Potentilla anglica	n = 28	2n = 56
5 Potentilla norvegica	n = 35	2n = 70
6 Potentilla tabernaemontani	n = 42	2n = 84

The following statements suggest various ways in which species 6 could have arisen by polyploidy from one or more of the other species. Which of these is *not* possible?
A Spindle failure and polyploidy occurring in a meristematic cell of species 3.
B The product of a cross between species 1 and 5 being crossed with species 3.
C A diploid gamete of species 5 fusing with a haploid gamete of species 2.
D The product of a cross between species 2 and 4 undergoing polyploidy.

37

9 The following diagram shows two chromosomes. The lettered regions represent genes.

```
   chromosome 1        chromosome 2
   P Q R S T U V W      E F G H
```

Which of the following would result if a translocation occurred between chromosomes 1 and 2?

A P Q R S W V U T   E F H G

B P Q R S   T U V W E F G H

C P Q S T U V W   E F H

D P Q R S T U V W V W
   E F G H G H

10 The following diagram shows the outcome of a certain type of chromosome mutation. The lettered regions indicate the positions of six marker genes.

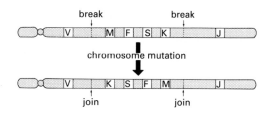

Which of the following diagrams best represents this mutated chromosome paired with its unaltered homologous partner during meiosis?

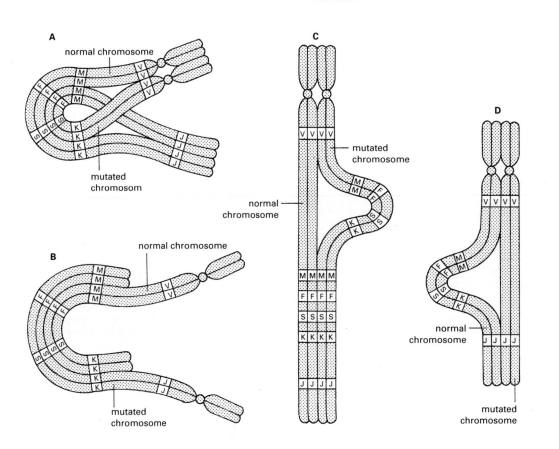

Items **11** and **12** refer to the following possible answers.
  A inversion   B deletion
  C insertion   D substitution

11 What name is given to the type of gene mutation where one incorrect nucleotide occurs in place of the correct nucleotide in a DNA chain?

12 What name is given to the type of gene mutation illustrated in the following diagram?

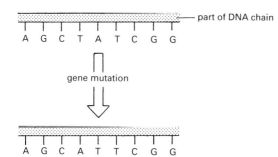

13 Which of the following is *not* a mutagenic agent?
  A nerve gas   B X-rays
  C low temperature   D ultraviolet rays

14 Neurofibrosis, a condition in which the human sufferer develops multiple brown lumps in the skin, is caused by a dominant mutant allele whose mutation rate is 100 per million gametes.
The chance of a new mutation occurring is therefore
  A 1 in 1000.   B 1 in 10 000.
  C 1 in 100 000.   D 1 in 1 000 000.

15 A mutation can occur in any part of the body at any time. In fruit flies it is most likely to spread through a population if it is present in a
  A wing cell.   B eye cell.
  C sperm cell   D salivary gland cell.

Test 17 Natural selection

1 Which of the following does *not* comprise part of the theory of evolution proposed by Darwin in his book *The Origin of Species*?
  A A struggle for existence occurs because organisms tend to produce more offspring than the environment will support.
  B Members of the same species are not identical but show variation in all characteristics.
  C Any beneficial change in an organism's phenotype is brought about by the direct action of the environment.
  D Those offspring whose phenotypes are less well suited to the environment are more likely to die before producing offspring.

2 Fossil evidence suggests that an animal only 0.4 m tall was the ancestor of the modern horse (1.5 m tall). Darwin would have explained this on the basis that over countless generations the animals that had survived were those that had
  A stretched their bodies in order to reach food high up in the trees.
  B grown longer legs as a direct result of constantly fleeing from predators.
  C inherited slightly taller stature than the other smaller members of the population.
  D elongated their necks in order to gain an elevated view of the surroundings.

Items **3**, **4**, **5**, **6**, **7** and **8** refer to the following information.
The peppered moth exists in two forms: the light-coloured variety and the dark (melanic) type. In an experiment, individuals of both types were marked on their underside with a dot of paint, and then some were released in a rural area and some were released in an industrial area.
Many of these marked moths were later recaptured, as shown in the table overleaf.

	rural area		industrial area	
	light moth	melanic moth	light moth	melanic moth
number of marked moths released	250	200	250	see item **5**
number of marked moths recaptured	40	see item **4**	45	162
percentage number of marked moths recaptured	16	4	18	54

3 Melanic moths occur as a result of
  A industrial pollution. B natural selection.
  C speciation. D mutation.

4 How many melanic moths were recaptured in the rural area?
  A 2 B 4 C 8 D 20

5 How many melanic moths were originally released in the industrial area?
  A 200 B 250 C 300 D 350

6 From the data in the table it is *not* valid to conclude that
  A in the rural area, light moths were four times more likely to survive than melanic moths.
  B a greater percentage number of both types of moth were recaptured in the industrial area compared with the rural area.
  C in the industrial area, melanic moths were three times more likely to survive than light-coloured moths.
  D the total percentage number of light moths recaptured in both areas exceeded the total percentage number of melanic moths recaptured.

7 The reason for marking each moth on its *underside* was to
  A treat all moths equally.
  B make the paint inconspicuous to predators.
  C avoid interfering with the moth's breathing system.
  D prevent the paint from damaging the moth's wings.

8 The melanic moth enjoys a selective advantage in an industrial area because
  A predators fail to notice it against a sooty background.
  B there is no competition since the light form is killed by pollution.
  C predators ignore it because it is dirty and noxious to eat.
  D it is easily seen against light-coloured tree trunks.

9 The following diagram shows the outcome of a cross between two sufferers of sickle cell trait (where H = allele for normal haemoglobin and S = allele for haemoglobin S).

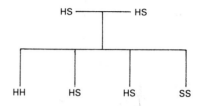

With respect to survival of the offspring, which of the following would be most likely?

	% number of survivors	
	population living in malarial area	population living in non malarial area
A	25	75
B	50	75
C	25	100
D	50	100

10 A higher proportion of African Negroes suffer from sickle cell trait than American Negroes because
  A such African Negroes enjoy a selective advantage.
  B American Negroes are outnumbered by white people.
  C only the African Negroes are resistant to malaria.
  D more American Negroes die of heart failure.

11 A mutant allele is most likely to give the mutant form of the organism a selective advantage if the
  A environment remains stable.
  B environment undergoes change.
  C mutant organism is heterozygous.
  D mutant organism population is large.

Items **12**, **13** and **14** refer to the following graph which shows the effect of an antibiotic on two strains of a species of bacterium.

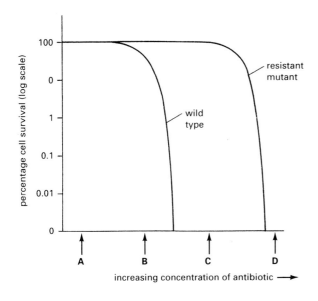

12 To which concentration of antibiotic are both strains of the bacterium sensitive?

13 Which concentration of antibiotic affects neither strain of bacterium?

14 Which concentration of antibiotic would be suitable for use in selecting the resistant mutant strain?

15 The viral disease, myxomatosis, was deliberately introduced into Australia in the early 1950s in an attempt to control rabbit populations.
The following table shows the results from an investigation using rabbits selected each year from wild populations and inoculated with the original disease-causing strain of virus.

year	% population suffering fatal symptoms
1952	93
1953	95
1954	93
1955	61
1956	75
1957	54

These results support the theory that
A over the years an increased number of genetically resistant rabbits survived.
B natural selection occurred between 1955 and 1957 with a peak in 1956.
C the virus which caused myxomatosis underwent a mutation each year.
D rabbits acquired an immunity to the disease in 1956 only.

16 Certain species of grass are able to tolerate high concentrations of copper in the soil. An analysis was made of the grass plants at sites P, Q and R in the diagram below.

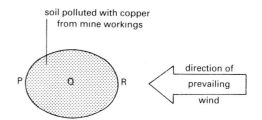

Which of the following correctly matches the three sites with the type(s) of grass plant that they would possess?

	all non-tolerant to copper	some tolerant, some non-tolerant	all tolerant to copper
A	P	Q	R
B	R	P	Q
C	R	Q	P
D	P	R	Q

17 Which of the following is *not* an example of natural selection in action?
A Emergence of rats which thrive on warfarin rat poison.
B Development of pedigree strains of Rottweiler dogs.
C Resistance of certain types of bacteria to penicillin.
D Survival of mutant headlice treated with insecticide.

Items **18**, **19** and **20** refer to the following information and diagrams of land snails.
The shell of the land snail shows variation in both colour and banding pattern. In order to construct a 5-figure banding formula, bands are numbered from the top of the largest whorl, as shown below. 0 is used to represent the absence of a band and square brackets indicate the fusion of two bands.

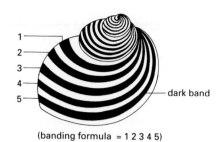

(banding formula = 1 2 3 4 5)

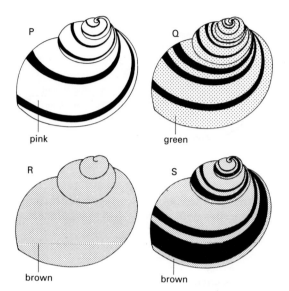

18 Shell S would have the banding formula
   A 030[45].   B 03045.
   C 02045.   D 003[45].

19 Thrushes (which have good colour vision) smash the shells of land snails against stones (anvils) in order to feed on the soft inner body.
   If snail types P, Q, R and S began in equal numbers in a habitat of grassland, which would be most likely to enjoy the greatest selective advantage?
   A P   B Q   C R   D S

20 A survey of broken shells collected from thrush anvils amongst dead beech leaves in a woodland area was carried out.
   Predict which of the following sets of results was obtained.

| | % number of broken shells of each type ||||
	P	Q	R	S
A	13	33	1	5
B	11	1	34	6
C	5	1	14	32
D	6	21	20	5

Test 18 Selective breeding of animals and plants

1 Which of the following statements is *false*?
   A A hybrid is the result of a cross between genetically dissimilar parents.
   B A hybrid is often stronger or better in some way than its parents.
   C A hybrid formed from two different species is sterile because its chromosomes fail to pair at meiosis.
   D A hybrid tends to be homozygous at many loci as a result of many generations of inbreeding.

2 In order to produce a supply of hybrids showing genetic uniformity, horticulturalists often maintain two different true-breeding parental lines of a species of bedding plant.
   The hybrids cannot be used as the parents of the next generation because
   A they are heterozygous and therefore not true-breeding.
   B hybrid vigour cannot be passed on to the next generation.
   C a high mutation rate occurs amongst hybrid gametes.
   D hybrids of annual plants always form sterile seeds.

3 The following cross involves two varieties of

the same species which have become homozygous as a result of many generations of inbreeding.

$$MMnnppQQ \times mmNNPPqq$$

The offspring from this cross would have the genotype

A MMnnPPqq.
B MmNnPpQq.
C mmNnPPQq.
D MMNNPPQQ.

4 Which of the following diagrams best represents selective breeding in the *Brassica* group of plants?

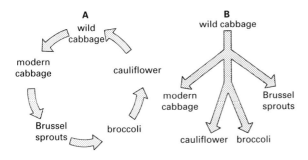

5 The following graph shows the effects of selective breeding on the milk yield and percentage butterfat produced by a certain breed of cattle over a thirty year period.

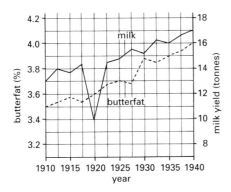

Which of the following shows the increases which occurred between 1910 and 1935?

	milk yield (tonnes)	butterfat %
A	0.4	3.0
B	0.5	4.0
C	3.0	0.4
D	4.0	0.5

6 Table 1 below shows the outcome of selfing four breeds of cattle (Q, R, S and T). Table 2 shows the outcome of hybridisation crosses involving the four breeds of cattle.

parents	average live weight of offspring at 18 months (kg)
Q × Q	300
R × R	350
S × S	250
T × T	300

Table 1

parents	average live weight of offspring at 18 months (kg)
Q × R	320
R × S	310
Q × S	280
S × T	290

Table 2

Which of the following crosses shows *negative* heterosis (i.e. offspring which are poorer than the mean of the two parents)?

A Q × R   B R × S   C Q × S   D S × T

Items **7**, **8** and **9** refer to the following information and the graphs that follow at the top of the next page.

Several types of artificial selection can be employed to alter a crop plant's quantitative characteristics.

1 *Disruptive* selection is practised when a crop is being developed for two different markets (e.g. barley with a low nitrogen content for brewing and barley with a high nitrogen content for the livestock feed).

2 *Stabilising* selection is used for maintaining uniformity (e.g. crop height to suit harvesting machinery).

3 *Directional* selection is practised if increase in yield per plant is required.

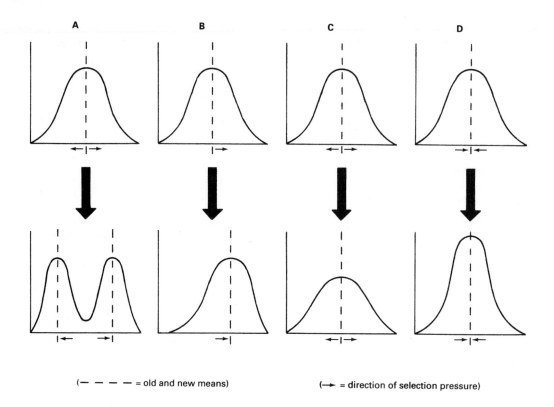

(— — — — = old and new means)   (→ = direction of selection pressure)

7 Which of the sets of graphs shown above represents disruptive selection?

8 Which set of graphs represents stabilising selection?

9 Which set of graphs represents directional selection?

10 P_1 in the cross shown opposite represents a cultivated variety of plant which contains the genes for many desirable characteristics. However, P_1 is susceptible to a particular virus.
   P_2 is a wild variety of the same species which possesses the gene for resistance to the virus.
   Following the first cross, 50% of the genetic material (including the resistance gene) received by P_3 comes from the wild parent. In order to dilute this unwanted wild genetic contribution, but retain the resistance gene, a series of crosses against P_1 is carried out. The first is shown in the diagram and results in the formation of P_4.

How many more crosses will need to be carried out to reduce the wild genetic material inherited by offspring to less than 5%?
A 1   **B** 2   **C** 3   **D** 4

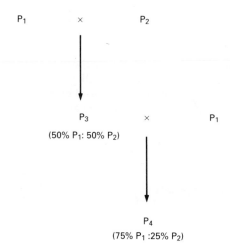

44

Test 19 Loss of genetic diversity

1 Which of the following statements is *false*?
   A Inbreeding increases the chance of individuals arising that are double recessive for an inferior allele.
   B Inbreeding depression often results from hybridisation between unrelated species.
   C Inbreeding results in loss of genetic diversity amongst members of a domesticated variety.
   D Inbreeding promotes the retention of desirable characteristics in a variety from generation to generation.

2 A species is most susceptible to loss of genetic diversity when it is
   A small in number and widespread.
   B small in number and restricted to one area.
   C numerous and widespread.
   D numerous and restricted to one area.

3 Pedigree dogs are produced by mating members of the same breed with one another. This often results in the production of offspring suffering conditions which affect their fitness. For example, bulldogs have problems with their breathing and labradors are prone to arthritis.
   This phenomenon is known as
   A hybrid depression.
   B natural selection.
   C inbreeding depression.
   D heterozygote selection.

Items 4 and 5 refer to the following diagram which charts the effect of repeated self-pollination on heterozygosity in a variety of flowering plant.

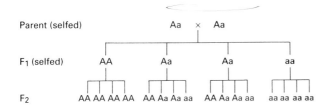

4 What percentage of the F_2 generation are heterozygous?
   A 4   B 12   C 25   D 50

5 If the pattern of selfing were repeated, by which generation would there be less than 1% of heterozygotes remaining in the population?
   A F_4 B F_5 C F_6 D F_7

Items 6, 7 and 8 refer to the following bar chart. The chart shows eight crop plants which provided the world with 75% of its food during one year in the 1980s.

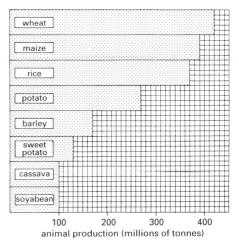

6 What was the *total* production of food (in millions of tonnes) obtained that year from *all* plant species in the world?
   A 1462.5   B 1950.0   C 2500.0   D 2600.0

7 The percentage of the world's food production derived that year from sweet potato was
   A 5.0%.   B 6.6%.   C 10.0%.   D 13.0%.

8 Which plant *alone* provided the world with 15% of its needs?
   A wheat   B maize   C rice   D potato

9 About eleven regions of the world are known to still possess wild relatives of today's domesticated crop plants which are living in their natural habitats.
   These regions are called
   A gene banks.
   B botanical gardens.
   C centres of diversity.
   D rare-breed farms.

10 The conditions required for successful storage of live material in a germ cell bank are

	humidity level	temperature
A	5%	−20°C
B	5%	20°C
C	95%	−20°C
D	95%	20°C

Test 20 Genetic engineering

1 The chromosomes shown in the following diagram have been stained and therefore show characteristic banding patterns. Which chromosome is abnormal in that is has undergone duplication of part of its genetic material?

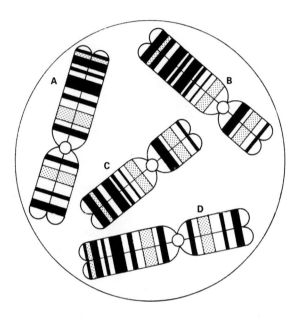

Items **2**, **3** and **4** refer to the following list of procedural steps employed during recombinant DNA technology.
 1 host cell allowed to multiply
 2 required DNA fragment cut out of appropriate chromosome
 3 duplicate plasmids formed which express 'foreign' gene
 4 plasmid extracted from bacterium and opened up
 5 recombinant plasmid inserted into bacterial host cell
 6 DNA fragment sealed into plasmid

2 The correct order in which these steps would be carried out is
 A 2,4,6,5,1,3. **B** 4,6,2,5,3,1.
 C 2,4,5,6,3,1. **D** 4,6,2,5,1,3.

3 The enzyme endonuclease would be employed during stages
 A 2 and 6.   **B** 2 and 4.
 C 4 and 5.   **D** 4 and 6.

4 The enzyme ligase would be employed at stage
 A 2.   **B** 4.   **C** 5.   **D** 6.

5 J, K, L and M are four genes known to be located on the same chromosome. Testcrosses between heterozygotes and homozygous recessives are found to give the results shown in the following table.

cross	% recombinants produced
JjKk × jjkk	18
Ll × llmm	38
MmJj × mmjj	24
LlKk × llkk	4

Which of the diagrams below best represents a linkage map of these genes on their chromosome?

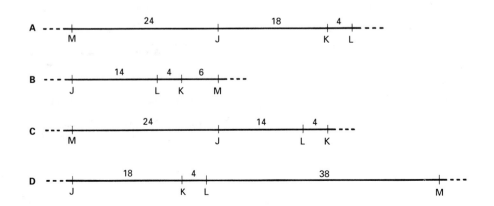

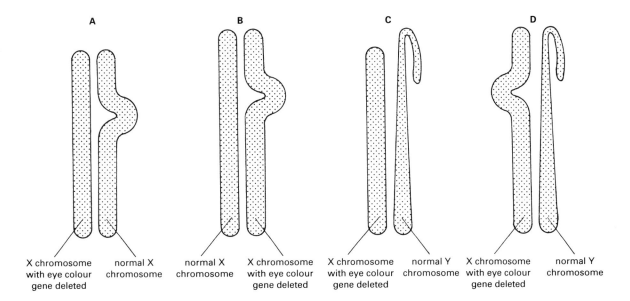

Items **6** and **7** refer to the following information.

To find the location on the X chromosome of the sex-linked gene for red/white eye colour in *Drosophila*, red-eyed males are exposed to radiation to induce mutations, and then crossed with white-eyed females.

6 The effect of this mutation is found to appear first in the phenotype of a few
  A males in the F_1 generation.
  B males in the F_2 generation.
  C females in the F_1 generation.
  D females in the F_2 generation.

7 Examination of sex chromosomes from the salivary glands of female larvae produced by the mutant strain, shows some of them to be unusual in that a deletion has occurred at the locus of the red/white eye colour gene.
  Which of the diagrams at the top of the page best shows the appearance these chromosomes would have?

Items **8** and **9** refer to the following information.
Restriction endonuclease enzymes do not cut DNA at random but recognise particular sequences of bases.

8 One enzyme has the following recognition sequence:

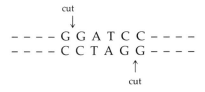

The diagram below shows a piece of DNA about to be cut.

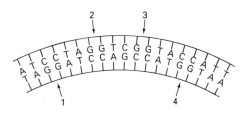

At which pair of numbered sites would the enzyme make its cuts?
A 1 and 2   **B** 3 and 4
C 1 and 3   **D** 2 and 4

9 The diagram below shows a different piece of DNA about to be acted upon by a second enzyme with the recognition sequence:

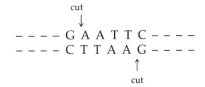

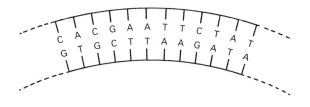

Which of the following diagrams shows the outcome of this enzyme's action?

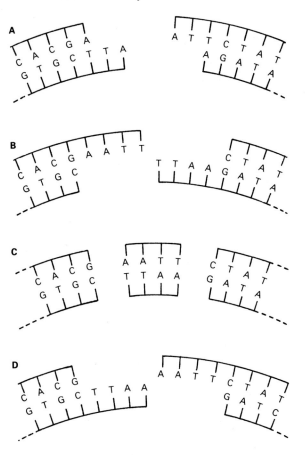

10 The list below gives the steps involved in bringing about somatic fusion between two species.
   1 isolated protoplasts induced to fuse by employing chemicals or electric currents
   2 callus treated with hormones to make it develop into a hybrid plant
   3 cell walls removed using the enzymes cellulase and pectinase
   4 unspecialised cells selected from two different species of plant
   5 hybrid protoplast allowed to divide forming mass of undifferentiated cells

The correct sequence in which these steps would be carried out is

A 3, 4, 1, 2, 5.   B 4, 3, 1, 5, 2.
C 3, 4, 1, 5, 2.   D 4, 3, 5, 1, 2.

Test 21 Speciation

1 Which of the following occurs during the process of speciation?
   A production of sterile offspring by interbreeding between two different species
   B alteration in a species' phenotype caused by environmental change
   C mass extinction of several species in a disturbed environment
   D formation of new species from existing ones

2 The sum total of the genes possessed by the members of an interbreeding population at a given time is known as the
   A gene frequency.   B gene code.
   C gene flow.   D gene pool.

3 Members of the same species
   A are reproductively isolated from one another.
   B share different gene pools.
   C possess the same chromosome complement.
   D are unable to interbreed and produce fertile offspring.

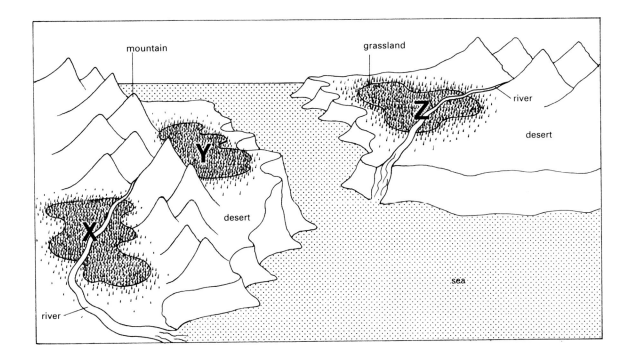

4 A population's gene pool was found to have remained unaltered. Which of the following conditions must have existed in the population?
A Mating had always been random.
B Genetic drift had often occurred.
C Inbreeding had been commonly practised.
D Certain alleles had enjoyed a selective advantage.

5 Which of the following does *not* alter a population's genetic equilibrium?
A high mutation rate of a gene
B immigration and emigration
C large size of breeding population
D artificial selection by humans

6 Three of the events that occur during speciation are
1 mutation.
2 natural selection.
3 isolation.
The correct order in which these occur is
A 3, 2, 1.   B 2, 1, 3.
C 3, 1, 2.   D 2, 3, 1.

7 Which of the following is *not* a reproductive barrier to speciation?
A non-correspondence of genital organs
B occurrence of polyploidy in a sterile hybrid
C inability of sperm to fertilise eggs
D failure of insects to pollinate flowers

8 Areas X, Y and Z in the diagram above represent three populations of a species of grassland-dwelling animal which is unable to fly or swim.
The barriers preventing interbreeding between populations X and Y, and Y and Z respectively are
A desert and sea.   B mountains and river.
C desert and river.   D mountains and sea.

Items **9**, **10** and **11** refer to the following diagram which shows the geographical distribution of five populations of a certain type of sea bird.

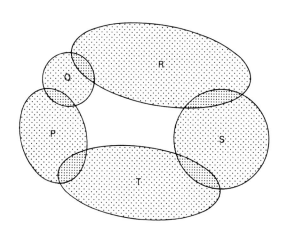

49

The table below shows the results of breeding experiments where + = successful interbreeding and − = unsuccessful interbreeding.

cross	result	cross	result
P×Q	+	Q×S	−
P×R	−	Q×T	−
P×S	−	R×S	+
P×T	−	R×T	−
Q×R	+	S×T	+

9 How many different species are present?
A 1   B 2   C 4   D 5

10 If population P became extinct, how many species would be present?
A 1   B 2   C 3   D 4

11 If, on the other hand, population S become extinct, how many species would be present?
A 1   B 2   C 3   D 4

12 About 440 million years ago, the world was dominated by warm seas and much of the land mass was submerged. 5 million years later, large ice sheets had formed, sea levels had dropped and vast areas of land had become exposed.
Which of the following types of organism would be most likely to suffer a wave of mass extinction under such circumstances?
A plankton living at the surface of the ocean
B deep sea invertebrate species
C marine animals adapted to warm shallows
D seaweeds native to cold waters

13 The process of extinction of species is decelerated by
A sudden climatic changes.
B overhunting by humans.
C habitat destruction.
D environmental conservation.

14 The following diagram shows the positions that the world's land masses are thought to have occupied millions of years ago before they gradually drifted apart.

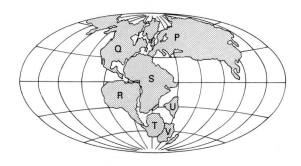

The spread of primitive marsupials to Australia is thought to have taken the route
A S→R→T→V.   B R→S→U→V.
C P→S→T→V.   D Q→S→U→V.

15 The land mass in the following map of the world is divided into six regions based on the types of mammal present in each region.

As a result of continental drift over millions of years, the mammals in
A regions 1 and 2 are very similar to one another.
B region 6 are very different to those in regions 1 and 2.
C regions 4 and 5 are very similar to one another.
D region 1 are very different to those in region 3.

Test 22 Adaptive radiation

1. Adaptive radiation is the evolution from
   A a single ancestral stock, of several convergent forms adapted to share an available ecological niche.
   B several divergent stocks, of a superior form adapted to fill an available ecological niche.
   C a single ancestral stock, of several divergent forms adapted to fill different ecological niches.
   D several divergent stocks, of several superior forms adapted to fill different ecological niches.

2. Scientists believe that over millions of years, Australian mammals have become very different from other mammals as a direct result of
   A evolving pouches in which to rear their young.
   B following their own course of evolution in isolation.
   C developing reproductive systems homologous to placentals.
   D evolving in climatically unique ecosystems.

3. Which of the following diagrams best represents the adaptive radiation of Darwin's finches?

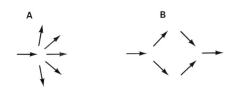

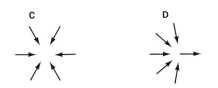

4. The following list includes three Australian marsupial mammals and three placental mammals which occupy similar ecological niches in other continents of the world.
   1 great red kangaroo
   2 flying squirrel
   3 sloth
   4 phalanger
   5 koala bear
   6 deer
   Which of the following correctly pairs each marsupial with the equivalent placental?
   A 1 & 6, 4 & 3, 5 & 2.
   B 1 & 3, 4 & 2, 5 & 6.
   C 1 & 2, 4 & 3, 5 & 6.
   D 1 & 6, 4 & 2, 5 & 3.

Items 5 and 6 refer to the following table.

	basic structure	common ancestor	function performed
A	same	no	different
B	different	no	same
C	same	yes	different
D	different	yes	same

5. Which description refers accurately to homologous structures?

6. Which description refers accurately to analogous structures?

7. Which of the following show divergent evolution?
   A eyes of locusts and blackbirds
   B skeletons of tortoises and lobsters
   C wings of cockroaches and bats
   D forelimbs of pigeons and dolphins

8. Which of the following comprise a pair of analogous structures?
   A proboscis of a butterfly and sting of a bee
   B incisors of a human and tusks of an elephant
   C jaws of a locust and incisors of a sheep
   D tusks of a walrus and canines of a horse

9. Which of the following are examples of the evolution of homologous structures?
   A legs of spider and crocodile
   B canine teeth of wolf and gorilla
   C hind limbs of tree frog and locust
   D shells of snail and turtle

10 The following table refers to five types of fruit and the structure which each possesses to effect seed dispersal.
Which group of fruits *all* possess structures which are homologous?
A 2, 3 and 4   B 1, 2 and 5
C 1, 3 and 5   D 3, 4 and 5

	fruit	dispersal structure and origin	mode of dispersal
1	pear	juicy edible coat formed mainly from receptacle of flower	animal (internal)
2	peapod	dry coat formed from ovary wall	mechanical
3	cherry	juicy edible coat formed from ovary wall	animal (internal)
4	ash	dry wing-like extension formed from ovary wall	wind
5	apple	juicy edible coat formed mainly from receptacle of flower	animal (internal)

Test 23 Growth differences between plants and animals

1 Meristematic cells are
   A found only at root and shoot tips in plants.
   B undifferentiated and capable of dividing repeatedly.
   C widely distributed throughout a developing animal's body.
   D used to transport materials in a plant's stem.

Items **2** and **3** refer to the following information.
   During growth at a plant apex, each cell undergoes the following processes.
   1 differentiation
   2 elongation
   3 division
   4 vacuolation

2 The order in which these occur is
   A 3, 2, 4, 1.   B 2, 3, 4, 1.
   C 3, 2, 1, 4.   D 2, 3, 1, 4.

3 Increase in the length of a plant depends on the occurrence of *both*
   A 1 and 2.   B 1 and 3.
   C 2 and 3.   D 1 and 4.

Items **4** and **5** refer to the following diagram of a root tip.

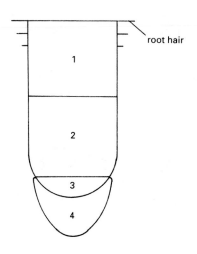

4 The diagram that follows shows a certain type of cell undergoing change during its development.

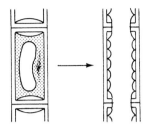

In which region of the root would this process of change occur?
A 1   **B** 2   **C** 3   **D** 4

5 The following diagram shows the appearance of two cells (X and Y) at different stages of their development.

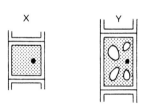

In which regions of the root would such cells be found?

	X	Y
A	2	1
B	2	3
C	3	2
D	4	3

Items **6** and **7** refer to the following diagram which shows a transverse section through a shoot apex.

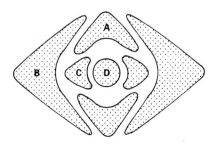

6 Which structure gives rise to leaf primordia?

7 In which structure will mitosis first come to a halt?

8 The diagram that follows shows a longitudinal section through a shoot apex.

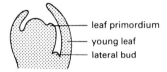

Which of the diagrams below shows the correct appearance of this shoot apex at the formation of the next leaf primordium?

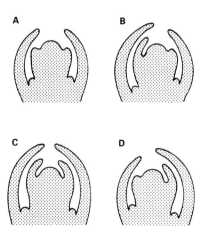

9 The following diagram shows the shoot apex of a young plant marked at regular intervals with waterproof ink.

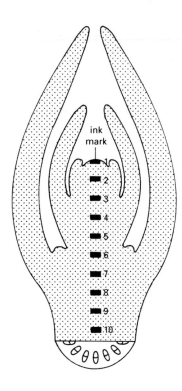

Items **10**, **11** and **12** refer to the piece of wood, cut from a felled tree, shown in the diagram below.

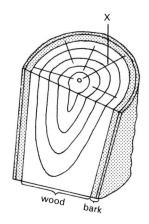

10 The age of the tree in years is
    A 4.   **B** 5.   **C** 6.   **D** 7.

11 Structure X is
    A a lenticel.
    B a medullary ray.
    C the secondary cortex.
    D a group of cork cells.

12 The function of X is
    A gaseous exchange with the external environment.
    B production of spring wood each year.
    C lateral transport of water and minerals.
    D prevention of cracking as the stem becomes thicker.

13 The following diagram shows a transverse section of a woody stem.

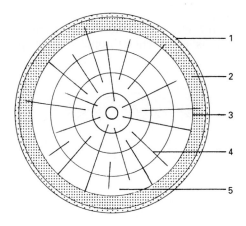

Which of the diagrams below best represents the marked region of this shoot apex after several days of growth?

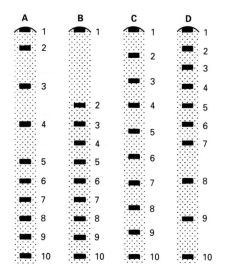

In which numbered regions of the diagram would parenchyma, xylem vessels and cambium be found?

	parenchyma	xylem vessels	cambium
A	4	5	3
B	1	3	2
C	5	2	1
D	3	4	5

14 From which region of the following woody stem has sample X been taken?

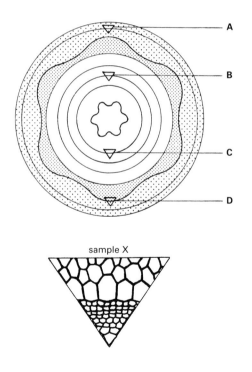

15 A certain locality's annual rainfall figures for four years are shown in the following table.

year	annual rainfall (mm)
1983	813
1984	978
1985	826
1986	672

A tree growing in this region was cut down in December 1986. Which of the following diagrams best represents a transverse section of this tree?

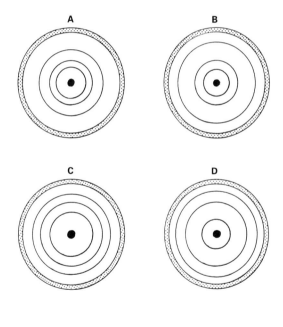

16 Cork cambium forms
  A both secondary cortex and cork cells to its inside.
  B both secondary cortex and cork cells to its outside.
  C secondary cortex to its outside and cork cells to its inside.
  D secondary cortex to its inside and cork cells to its outside.

17 Which longitudinal cut through the tree shown (in transverse section) in the following diagram would produce the grain pattern shown below?

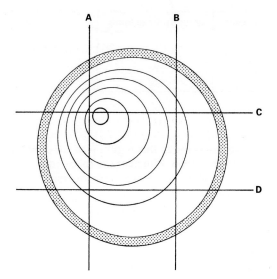

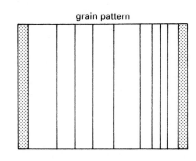

18 All of the transverse sections shown in the following diagram are from one plant.

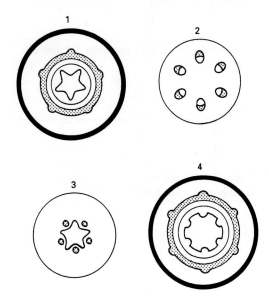

Water entering the plant at a root hair would pass through these in the order
A 1,3,4,2.   B 1,4,3,2.
C 3,1,4,2.   D 3,2,1,4.

19 Which of the following tissues *can* be regenerated by the human body?
A muscle   B heart   C bone   D brain

20 Which of the following cell types *cannot* be regenerated by the human body?
A blood   B skin   C liver   D lung

Test 24 Growth patterns

1. Which of the following is the most reliable indicator of growth in an annual plant (e.g. broad bean)?
   A fresh mass
   B dry mass
   C shoot length
   D root length

Items **2**, **3** and **4** refer to the following diagram which shows a typical S-shaped growth curve.

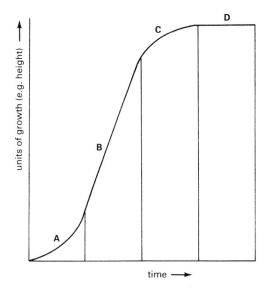

2. Which region of the graph represents the period of decelerating growth?

3. Which region of the graph represents the period of accelerating growth?

4. Which region of the graph represents the period of no growth?

Items **5**, **6** and **7** refer to the following four graphs.

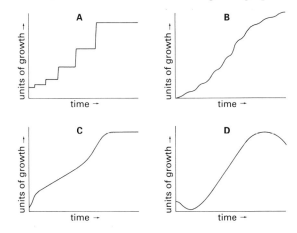

5. Which graph best represents the growth curve of an oak tree's height?

6. Which graph best represents the growth curve of a cockroach's body length?

7. Which graph best represents the growth curve of a human being's body mass?

8. The following table compares the growth of a rose bush with that of a gerbil. Which pair of statements is *inaccurate*?

	gerbil	rose bush
A	increase in size stops on reaching adulthood	increase in size continues throughout life
B	growth occurs all over body	growth occurs only at meristems
C	regenerative powers are very limited	regenerative powers are fairly extensive
D	does not show an increase in dry mass	does show an increase in dry mass

Items **9** and **10** refer to the following graph, which shows the rate of increase in height of boys and girls between the ages of six months and eighteen years (based on data from a large population).

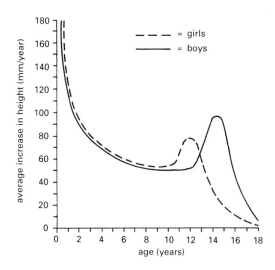

9. Between which ages did the boys gain height at their fastest rate?
   A 10–11
   B 11–12
   C 12–13
   D 13–14

10 On average, the girls showed an annual gain in height of 80 mm at ages
   A 3 and 12. B 3 and 13.
   C 3 and 15. D 13 and 15.

11 The following graph charts the growth, in length, of a human foetus before birth.

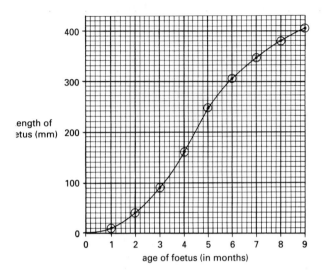

   What was the average rate of growth of the foetus, in mm/month, during the final four months of the pregnancy?
   A 25   B 40   C 160   D 405

12 The data in the table below refers to a small mammal.

age (days)	mass (g)	gain in mass since last weighing (g)	time interval between weighings (days)	average daily gain in mass (g)
3	3	–	–	–
6	4	1	3	0.33
12	10	6	6	1.00
28	30	20	16	box Y
38	54	box X	10	2.40

The figures in boxes X and Y should be

	X	Y
A	27	1.25
B	27	1.33
C	24	1.25
D	24	1.33

13 In an insect's life cycle, each stage which occurs between two moults (ecdyses) is called an instar. A locust passes through several instar stages before reaching its final size.
   The graph below shows the effect of temperature on rate of development in the locust.

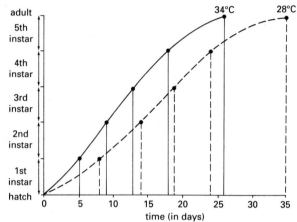

Which of the following is *not* true?
A Insects at the lower temperature take 9 days longer to reach the adult stage.
B Insects at the higher temperature have reached the fifth instar after 18 days.
C At the lower temperature insects take 3 days longer to reach the second instar.
D At both temperatures insects have reached the fourth instar after 15 days.

Items **14** and **15** refer to the table below which gives the average height of human males at different ages.

age (in years)	height (in mm)
birth	506
2	875
4	1034
6	1175
8	1300
10	1403
12	1496
14	1627
16	1716
18	1745

14 Maximum increase in height occurs between
   A birth and 2 years.   B 4 and 6 years.
   C 14 and 16 years.   D 16 and 18 years.

15 A histogram drawn from this data, plotting increase in height against age, would have the appearance:

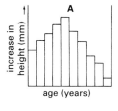
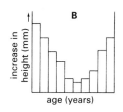
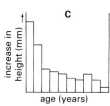
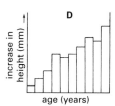

Test 25 Genetic control

1 Each cell in a multicellular organism contains all the genes necessary for the construction of
   A that one cell only.
   B all cells of that type of tissue.
   C all types of that cell.
   D the whole organism.

2 Human gastric glands contain pepsin-secreting cells. Within each of these differentiated cells
   A all genes continue to operate.
   B certain genes no longer operate.
   C certain genes are no longer present.
   D none of the genes operate.

3 Which of the following structures *both* contain pools of undifferentiated ('stem') cells which give rise to human blood cells?
   A red bone marrow and spleen
   B spleen and heart
   C heart and lymph glands
   D lymph glands and liver

4 In a monocyte, some of the genes that are 'switched on' are those required for the formation of
   A antibodies.
   B digestive enzymes.
   C haemoglobin.
   D blood-clotting chemicals.

5 In a lymphocyte, some of the genes that are 'switched on' are those required for the formation of
   A antibodies.
   B digestive enzymes.
   C haemoglobin.
   D blood-clotting chemicals.

6 In the parenchyma cells which make up the fruit wall of a ripening tomato, the genes that are 'switched on' would be those required for the formation of
   A chloroplasts containing red pigment.
   B chloroplasts containing green pigment.
   C chromoplasts containing red pigment.
   D chromoplasts containing green pigment.

7 In a horse chestnut tree, the cells in a leaf primordium do not give rise to a stem or root because such cells
   A only possess the genes required to form leaves.
   B are located at an elevated position on the plant.
   C have only the genes for leaf formation switched on.
   D suffer loss of root and stem genes during cell division.

8 In a polygenic inheritance pattern,
   A one phenotypic characteristic is controlled by several genes situated at different loci.
   B several different expressions of a phenotypic characteristic are controlled by one gene.
   C one phenotypic characteristic is controlled by several alleles of one gene.
   D several different expressions of a phenotypic characteristic are controlled by a group of genes at one locus.

Items 9, 10 and 11 are based on the assumption that human skin colour is determined by two independent genes whose effects are additive. Each gene has two alleles, as summarised in the following table.

	dominant allele	recessive allele
gene 1	M^1 (codes for 25% of maximum possible amount of melanin)	m^1 (does not code for melanin)
gene 2	M^2 (codes for 25% of maximum possible amount of melanin)	m^2 (does not code for melanin)

9 If several couples who all possess the genotype $M^1m^1M^2m^2$ produce a total of 32 offspring, then the number of these offspring expected statistically to have skin which is a darker colour than their parents would be
   A 2.   B 5.   C 8.   D 10.

10 Amongst the 32 offspring referred to in item 9, the number of different tones of skin colour that would be expected statistically to occur would be
   A 2.   B 5.   C 9.   D 16.

11 Which of the following crosses could *not* produce white-skinned offspring?
   A $M^1m^1M^2m^2 \times m^1m^1M^2m^2$
   B $M^1m^1m^2m^2 \times m^1m^1M^2M^2$
   C $M^1m^1m^2m^2 \times M^1m^1M^2m^2$
   D $M^1m^1M^2m^2 \times M^1m^1M^2m^2$

12 The percentage amount of light reflected by human skin exposed to red light of a certain wavelength varies depending on skin colour.
    The following graph charts this effect with respect to pure bred black and pure bred white people.

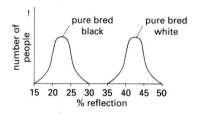

If a large number of these black people all married white people, and their children then intermarried, which of the following distributions would represent the next generation?

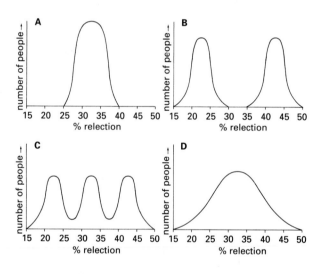

13 An iris of the human eye which is rich in melanin is
   A dark brown and absorbs much light.
   B dark brown and reflects much light.
   C pale blue and absorbs much light.
   D pale blue and reflects much light.

Items 14 and 15 refer to the following information.
    The colour of the iris in the human eye is thought to be controlled by three independent genes whose dominant alleles (B^1, B^2 and B^3) code for melanin and have an additive effect. The recessive alleles (b^1, b^2 and b^3) fail to code for melanin. To possess brown eyes, a person must have at least three B-alleles in his or her genotype.

14 The chance of a brown-eyed child being produced by parents with the genotypes $B^1b^1b^2B^3b^3$ and $b^1b^1B^2b^2b^3b^3$ is
A 1 in 1. B 1 in 2.
C 1 in 4. D 1 in 8.

15 The genotype of the brown-eyed child referred to in item 14 would be
A $B^1b^1B^2B^2B^3b^3$. B $B^1b^1b^2b^2B^3b^3$.
C $B^1b^1B^2b^2B^3b^3$. D $B^1b^1B^2b^2b^3b^3$.

Test 26 Control of gene action

Items **1**, **2** and **3** refer to the diagram directly below which shows a possible arrangement of the genes involved in the induction of the enzyme β-galactosidase in *Escherichia coli*.

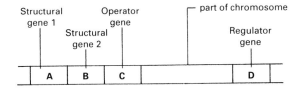

1 To which gene would the repressor molecule become attached in the absence of lactose?

2 Which gene produced the repressor molecule?

3 In this system the inducer molecule is called
A glucose. B galactose.
C lactose. D β-galactosidase.

Items **4**, **5** and **6** refer to the following information, the diagram below and table of possible answers overleaf.

Tryptophan is an amino acid needed for the synthesis of proteins. Situation 1 in the diagram shows a set of circumstances where certain genes remain 'switched on' in a cell of *Escherichia coli*.

Under different circumstances, the series of events shown in situation 2 is thought to occur. This brings about the repression of the synthesis of an enzyme.

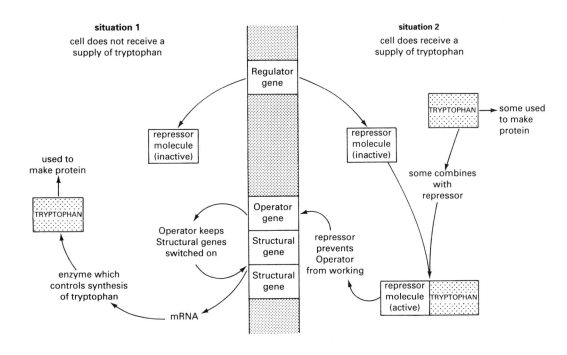

	Regulator gene	Operator gene	Structural gene
A	on	off	off
B	off	off	off
C	off	on	on
D	on	on	on

4 Which answer refers to the state of the genes in a cell of *Escherichia coli* grown on nutrient agar lacking tryptophan?

5 Which answer refers to the state of the genes in a cell of *Escherichia coli* cultured in nutrient broth containing tryptophan?

6 Which answer refers to the situation that would arise if the supply of tryptophan referred to in item **5** ran out?

7 The following statements describe possible benefits to a cell from being able to switch genes on and off as required. Which statement is *incorrect*?
  A A steady state is maintained despite fluctuations in supply of materials.
  B A constant supply of energy is made available for diffusion of ions to occur.
  C Needless wastage of resources such as amino acids is prevented.
  D Efficient use is made of available energy during cell metabolism.

Items **8**, **9** and **10** refer to the diagram at the bottom of the page which represents the last four stages in a metabolic pathway in the fungus *Neurospora*.

8 A mutant strain of the fungus is found to accumulate compound Q as a result of its metabolism. The gene which has undergone a mutation in this strain is
  A 1.   **B** 2.   **C** 3.   **D** 4.

9 Wild type *Neurospora* can grow on minimal medium (sucrose, mineral salts and one vitamin) but mutant strains suffering metabolic blocks are unable to do so.
  In an experiment the mutant strain referred to in item **8** was subcultured onto the following plates.
  Plate P = minimal medium + substance P
  Plate Q = minimal medium + substance Q
  Plate R = minimal medium + substance R
  Plate S = minimal medium + substance S
  It would grow successfully on *both* plates
  A P and Q.   **B** Q and R.
  C R and S.   **D** S and P.

10 A different mutant strain was found to grow successfully on plate S (minimal medium + substance S), but on no other.
  The enzyme that this strain of *Neurospora* fails to make is
  A 1.   **B** 2.   **C** 3.   **D** 4.

11 The diagram below shows a simplified version of a biochemical pathway that occurs during cell metabolism in normal humans.
  Which gene has undergone a mutation in a sufferer of phenylketonuria?

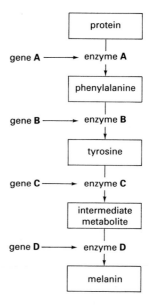

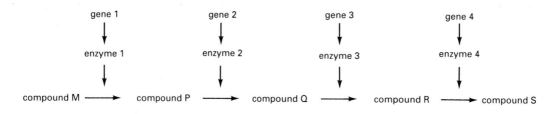

12 Which of the following statements about phenylketonuria is *false*?
  A It occurs in approximately 1 in 10 000 of the British population.
  B Sufferers completely lack pigmentation of skin, hair and eye irises.
  C It leads to the formation of phenylpyruvic acid which is excreted in urine.
  D Sufferers accumulate toxins which affect the metabolism of brain cells.

13 Which of the following are *both* gonadotrophic hormones?
  A follicle-stimulating hormone and luteinising hormone
  B luteinising hormone and progesterone
  C progesterone and oestrogen
  D oestrogen and follicle-stimulating hormone

14 Which of the following statements is *correct*?
  A Hypothalamus cells release gonadotrophic hormones which stimulate pituitary cells to make releaser hormone.
  B Hypothalamus cells make releaser hormone which stimulates pituitary cells to release gonadotrophic hormones.
  C Pituitary cells release gonadotrophic hormones which stimulate hypothalamus cells to make releaser hormone.
  D Pituitary cells make releaser hormone which stimulates hypothalamus cells to release gonadotrophic hormones.

15 The following statements refer to some of the events which occur at puberty in human males, as facial hair begins to develop.
  1 Testosterone and receptor form complex which enters target cell.
  2 Gonadotrophic hormone stimulates sex organs to release testosterone into bloodstream.
  3 Sex hormone enters nucleus and gene becomes switched on.
  4 Gene transcribes mRNA needed to code for hair protein.
  5 Sex hormone meets receptor on membrane of target cell in skin.

The correct order in which these occur is
  A 2, 5, 1, 3, 4.   B 5, 1, 2, 3, 4.
  C 2, 5, 3, 1, 4.   D 5, 3, 4, 1, 2.

Test 27 Hormonal influences on growth – part 1

1 The numbered structures in the diagram below represent two human endocrine glands. Which of the following indicates the correct sites of production of the three hormones?

	thyroxine	somato-trophin	thyroid-stimulating hormone
A	2	1	2
B	2	1	1
C	1	2	1
D	1	2	2

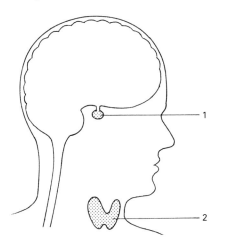

2 Acromegaly in the human body is caused by
  A underactivity of the pituitary gland during adolescence.
  B overactivity of the pituitary gland during adolescence.
  C underactivity of the pituitary gland during adulthood.
  D overactivity of the pituitary gland during adulthood.

Items **3** and **4** refer to the following bar graph.

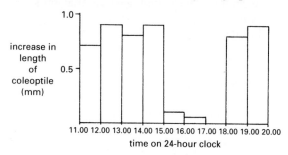

3 The oat coleoptile was decapitated at
  A 13.05 hours. **B** 14.05 hours.
  C 15.05 hours. **D** 16.05 hours.

4 Which of the following shows what was done to the coleoptile stump at 18.05 hours?

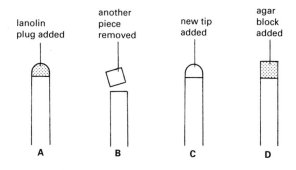

Items **5, 6, 7** and **8** refer to the following graph.

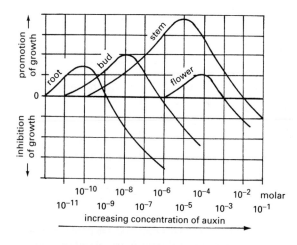

5 10^{-5} molar concentration of auxin causes maximum growth of the
  A stem. **B** root. **C** flower. **D** bud.

6 Which plant organ shows *least* promotion of growth over the whole range of auxin concentrations?
  A bud **B** flower **C** root **D** stem

7 Which molar concentration of auxin inhibits growth of roots yet promotes growth of both buds and stems?
  A 10^{-10} **B** 10^{-9} **C** 10^{-8} **D** 10^{-6}

8 Both promotion of stem growth and inhibition of flower growth occur between molar concentrations
  A 10^{-5} and 10^{-4}. **B** 10^{-4} and 10^{-3}.
  C 10^{-3} and 10^{-2}. **D** 10^{-2} and 10^{-1}.

9 Curvature of a shoot towards a light source is brought about by
  A increased plasticity of cell walls on the light side.
  B reduced concentration of growth substance on the dark side.
  C inhibition of cell division on the light side.
  D differential elongation of cells behind the shoot tip.

10 The following diagram shows four coleoptiles set up at the start of an experiment.

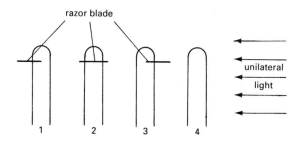

Which two coleoptiles will *both* bend towards the light source?
  A 1 and 2 **B** 1 and 4
  C 2 and 3 **D** 3 and 4

11 Agar blocks 1 and 2 were kept in the positions shown in the diagram below for several hours and then transferred onto two freshly cut coleoptiles.

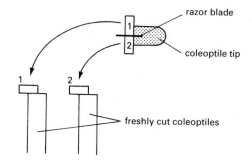

Which of the following would result after two days of growth?

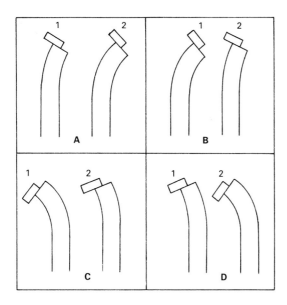

12 Gardeners do not have to worry about planting seeds upside down because roots, on emerging at germination, show
A negative phototropism.
B positive phototropism.
C negative geotropism.
D positive geotropism.

13 Tubes of cress seedlings were set up as shown in the following diagram.

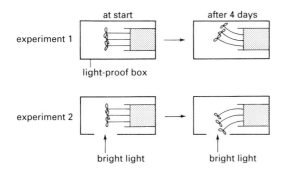

From the results of these experiments it can be concluded that
A negative phototropism is a more powerful response than positive geotropism.
B positive phototropism is a more powerful response than negative geotropism.
C positive geotropism is a more powerful response than negative phototropism.
D negative geotropism is a more powerful response than positive phototropism.

Items **14** and **15** refer to the experiment shown in the following diagram.

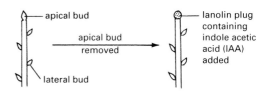

14 After two weeks the appearance of the shoot would be

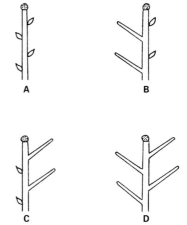

15 The most suitable control for this experiment would be

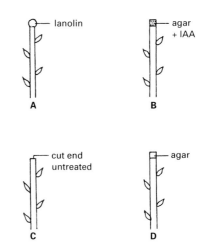

Test 28 Hormonal influences on growth – part 2

1 The diagram below shows a normal untreated dwarf pea plant.

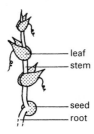

If its shoot had been treated with gibberellic acid during the early stages of germination, it would instead have the appearance

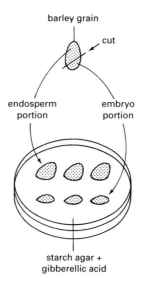

A   B   C   D

2 Assuming that the production of sufficient gibberellin to produce a tall pea plant is controlled by the dominant allele (G) of the gene for height, then the genotypes of a dwarf and a tall pea plant would be respectively
A g and G.   B g and Gg.
C gg and GG.   D Gg and GG.

Items 3, 4 and 5 refer to the following experiment, where soaked barley grains are cut and placed, cut surface down, on a plate of starch agar containing gibberellic acid.

3 Which of the following best represents the appearance of the plate 24 hours later when the grains are removed and the plate is flooded with iodine solution?

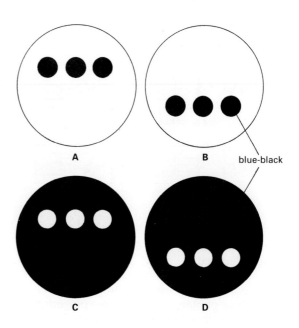

Items **4** and **5** refer, in addition, to the following diagram of the internal structure of a barley grain.

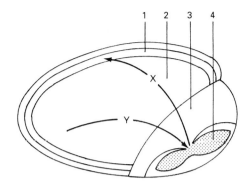

4 Which number indicates the aleurone layer?
  A 1   **B** 2   **C** 3   **D** 4

5 It is thought that in a germinating barley grain, gibberellic acid passes along route
  A X, and digests starchy endosperm to sugar in region 1.
  B X, and induces α-amylase production in region 1.
  C Y, and supplies sugar to the cells in region 4.
  D Y, and induces α-amylase production in region 4.

6 Which of the following effects is brought about by gibberellins but *not* by auxins?
  A maintenance of dormancy in lateral buds
  B inhibition of leaf abscission
  C promotion of phototropic responses
  D breaking of dormancy in leaf buds

7 Synthetic auxins are *not* used to
  A selectively kill broad-leaved garden weeds.
  B prevent destruction of rose bushes by aphids.
  C stimulate formation of adventitious roots.
  D induce parthenocarpy in unpollinated flowers.

8 Which of the following effects is brought about by *both* gibberellic acid and indole acetic acid?
  A reversal of genetic dwarfism
  B breaking of dormancy in seeds
  C promotion of cell elongation
  D induction of α-amylase in seed grains

Items **9** and **10** refer to the following graph, which shows the effect of the recommended concentration of a selective herbicide on a cereal crop and three different species of broad-leaved weed, at different stages of their development.

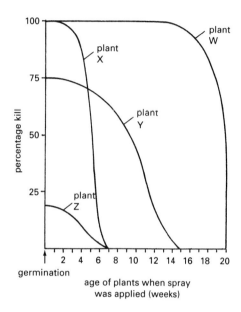

9 Which curve represents the cereal crop?
  A W   **B** X   **C** Y   **D** Z

10 In which of the following weeks would application of this concentration of the herbicide spray be *most* effective?
  A 4   **B** 7   **C** 15   **D** 20

Test 29 Effects of chemicals on growth

1 Which of the following is *not* required when setting up a water culture experiment (to determine the importance of individual chemical elements to a plant)?
   A distilled water containing one essential element
   B a glass tube to aerate the plant's roots
   C nitric acid to remove mineral traces from glassware
   D an opaque cover around the culture solution

Items **2**, **3** and **4** refer to the chemical elements given in the following list.
   1 potassium
   2 nitrogen
   3 magnesium
   4 phosphorus
   5 iron

2 Which element is an essential constituent of all proteins?
   A 1   B 2   C 3   D 4

3 The formation of chlorotic leaves is the result of a plant being deficient in *either*
   A 1 or 2.   B 2 or 3.
   C 3 or 4.   D 1 or 3.

4 Which element is required by a plant for the formation of ATP and nucleic acids?
   A 1   B 3   C 4   D 5

Items **5** and **6** refer to the experiment shown in the following diagram.

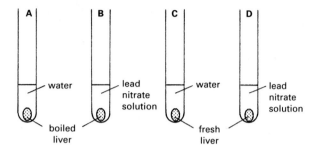

5 An equal volume of hydrogen peroxide (H_2O_2) was added to each of the test tubes. In which one did a reaction occur which released a froth of oxygen bubbles?

6 This experiment demonstrates that lead
   A promotes the action of the enzyme catalase.
   B inhibits the action of the enzyme catalase.
   C promotes the action of the enzyme hydrogenase.
   D inhibits the action of the enzyme hydrogenase.

Items **7** and **8** refer to the table below which gives the results of a survey on lead in drinking water.

		number of persons				
lead concentration in blood (mg/l)	lead concentration in drinking water (mg/l)	0.000 to 0.010	0.011 to 0.050	0.051 to 0.100	0.101 to 0.300	0.301 to above
	0.000 to 0.100	15	2	0	1	0
	0.101 to 0.200	27	10	9	11	3
	0.201 to 0.300	8	3	2	10	7
	0.301 to 0.400	0	0	1	3	4
	0.401 to 0.500	0	0	0	1	2
	0.501 and above	0	0	0	0	1

7 The most commonly consumed concentration of lead in drinking water, and the number of people found consuming it, are respectively

	lead concentration (mg/l)	number of persons
A	0.000 to 0.010	50
B	0.000 to 0.010	60
C	0.101 to 0.200	50
D	0.101 to 0.200	60

8 The highest concentration of lead found in blood, and the number of people possessing it, are respectively

	lead concentration (mg/l)	number of persons
A	0.301 and above	1
B	0.301 and above	17
C	0.501 and above	1
D	0.501 and above	17

9 Iron is *not* lost from the human body in
 A bile.
 B urine.
 C saliva.
 D dead skin cells.

10 Iron is an essential constituent of *both*
 A haemoglobin and cytochrome.
 B cytochrome and lymph.
 C lymph and plasma.
 D plasma and haemoglobin.

11 Calcium is *not* required by the human body for
 A clotting of blood.
 B formation of teeth.
 C contraction of muscle.
 D manufacture of bile.

12 The table below shows the calcium content (per 100 g) of several foodstuffs.

foodstuff	calcium content (mg/100 g)
almonds	250
blackcurrants	60
chocolate biscuits	130
cottage cheese	80
herring	100
ice-cream	140
milk	120
milk chocolate	250
oranges	40
sardines	400
white bread	100
yoghurt	140

The minimum recommended daily intake of calcium for a 16 year old is 600 mg. This could be achieved by eating
 A 100 g almonds, 200 g oranges and 200 g chocolate biscuits.
 B 100 g cottage cheese, 200 g ice cream and 400 g blackcurrants.
 C 100 g milk chocolate, 200 g herring, 100 g yoghurt.
 D 50 g sardines, 250 g white bread, 100 g milk.

13 In growing children, vitamin D is essential to promote
 A reabsorption of minerals from the bones.
 B storage of excess iron in the liver.
 C uptake of iodine by the thyroid gland.
 D absorption of calcium from the intestine.

14 Which of the following foods are *both* rich in vitamin D?
 A cod-liver oil and blackcurrants
 B blackcurrants and raw carrots
 C raw carrots and egg yolk
 D egg yolk and cod-liver oil

15 If a pregnant woman drinks alcohol in excess, the developing embryo is adversely affected because
 A it fails to receive an adequate oxygen supply.
 B its nervous system is artificially stimulated.
 C it receives excess vitamin B and zinc.
 D it is deprived of soluble carbohydrate.

Test 30 Effect of light on growth

1. The table below shows a comparison between two genetically identical plants, one grown in darkness, the other in light.
   Which paired statement is *incorrect*?

	in darkness	in light
A	leaves remain curled	leaves become expanded
B	weak internodes develop	strong internodes develop
C	leaves fail to make chlorophyll	leaves make chlorophyll
D	internodes remain short	elongated internodes develop

2. A short day plant will flower only when the continuous period of
   A light is above a critical level.
   B light is below a critical level.
   C darkness is below a critical level.
   D darkness is interrupted at a critical level.

3. Maryland Mammoth Tobacco is a short day plant. Its critical duration of darkness is 10 hours. Under which of the following conditions will it *not* flower?

4. When a certain species of flowering plant is exposed to artificially controlled periods of light and dark, it responds as shown in the following table.

dark period (hours)	light period (hours)	flowering (+) or no flowering (−)
10	12	+
11	11	−
11	12	+
12	11	−
12	12	+
13	11	−

   This species is a
   A long day plant, and the critical duration of light is 11 hours.
   B long day plant, and the critical duration of light is 12 hours.
   C short day plant, and the critical duration of darkness is 11 hours.
   D short day plant, and the critical duration of darkness is 12 hours.

5. Phytochrome exists in two forms, P_{660} and P_{730}. Which of the following conversions does *not* occur?

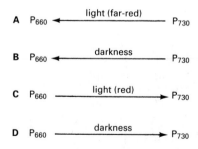

6. The photoperiodic stimulus is detected in a plant by the
   A leaves.   B buds.
   C flowers.   D shoot tips.

7. Which of the following diagrams correctly represents the events that occur in a short day plant?

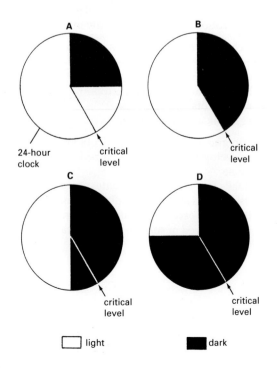

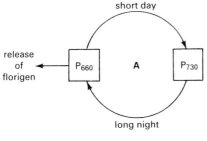
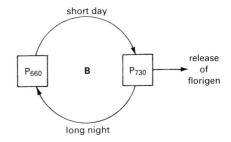
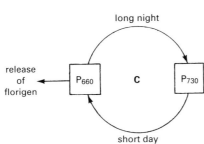
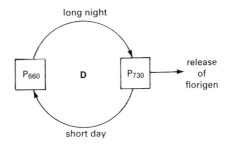

8 Cocklebur is a short day plant. After exposure to the treatments described in the following diagram for several days, plants X and Y were grafted together.

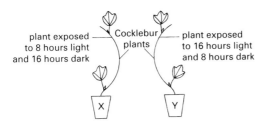

Which of the following would result on flowering?

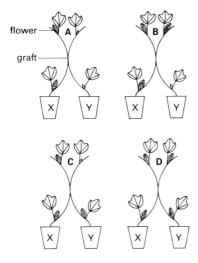

9 Sedum is a long day plant. It does not flower at the equator because
   A the nights are too dark.
   B there is great variation in daylength.
   C the days never become long enough.
   D it needs to be exposed to a period of frost.

10 The mammalian 'biological clock' is thought to be the
   A pineal gland which secretes melatonin in darkness.
   B hypothalamus which secretes melatonin in darkness.
   C pineal gland which secretes melatonin in light.
   D hypothalamus which secretes melatonin in light.

11 Some of the events which precede mating behaviour in sheep are
   1 secretion of gonadotrophic hormones
   2 arrival of shorter photoperiods in autumn
   3 secretion of sex hormones which induce mating behaviour
   4 stimulation of pituitary by releaser hormone
   5 stimulation of hypothalamus

   The order in which these events occur is
   A 2, 5, 1, 4, 3.   B 2, 5, 4, 1, 3.
   C 5, 2, 4, 3, 1.   D 5, 2, 4, 1, 3.

Items **12** and **13** refer to the following graph and table which refer to testis volume of the wood pigeon and changing photoperiod respectively.

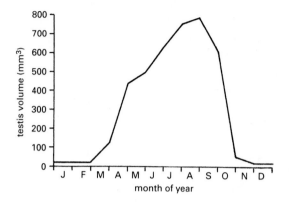

month of year	average photoperiod (hours)
J	8
F	10
M	12
A	14
M	16
J	18
J	17
A	15
S	13
O	11
N	9
D	7

12 If breeding behaviour and release of sperms do not occur until the testis has reached a volume of 300 mm³, then the critical photoperiod (in hours) is
   A 12.   **B** 14.   **C** 16.   **D** 18.

13 The breeding season of this bird is from
   A March to September.
   B April to September.
   C March to October.
   D April to October.

Items **14** and **15** refer to the table below which contains data obtained from wild populations of three species of bird living in the northern hemisphere.

species	latitude	maximum gonad size (mm³)
X	34°	400
X	45°	490
Y	47°	700
Y	52°	1500
Z	52°	450
Z	57°	955

14 Which of the following statements is *correct*?
   A Compared with X and Y, gonad size is greatest in species Z.
   B For each species, increase in latitude is accompanied by an increase in gonad size of at least 50%.
   C The higher the latitude, the greater the gonad size of all three species.
   D An increase in latitude from 34° to 57° is accompanied by an increase in gonad size of 555 mm³.

15 The trend shown by the data in the table can be accounted for by the fact that
   A breeding behaviour in birds at higher latitudes begins in winter.
   B breeding behaviour in birds at lower latitudes begins in winter.
   C summer daylengths are longer at higher latitudes.
   D summer daylengths are shorter at higher latitudes.

Test 31 Physiological homeostasis

1. Negative feedback control involves the following four stages.
   1. Effectors bring about corrective responses.
   2. A receptor detects a change in the internal environment.
   3. Deviation from the norm is counteracted.
   4. Nervous or hormonal messages are sent to effectors.

   The order in which these occur is
   A 2, 1, 4, 3. B 2, 4, 1, 3.
   C 2, 4, 3, 1. D 4, 2, 1, 3.

2. In humans, an increase in urine production occurs as a result of
   A a decrease in environmental temperature.
   B a decrease in water uptake.
   C an increase in salt uptake.
   D an increase in rate of sweating.

3. When a decrease in water concentration of the blood occurs, which of the following series of events brings about homeostatic control?

	ADH production	permeability of kidney tubules	volume of urine produced
A	increased	increased	decreased
B	increased	decreased	decreased
C	decreased	increased	decreased
D	increased	increased	increased

4. Set point is defined as the
   A point at which the body corrects any change in internal environment.
   B range within which the level of all factors must remain.
   C optimum position about which any one factor varies continuously.
   D point at which the body always maintains the internal environment.

5. The following diagram shows a simplified version of the homeostatic control of blood sugar level in the human body.
   Which of the following responses would occur in a human body as a direct result of eating a school lunch?
   A R, P and T   B S, P and T
   C R, Q and T   D R, P and U

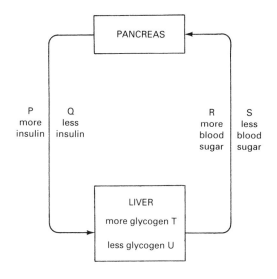

Items 6, 7 and 8 refer to the diagram below where the four lettered structures represent endocrine glands.

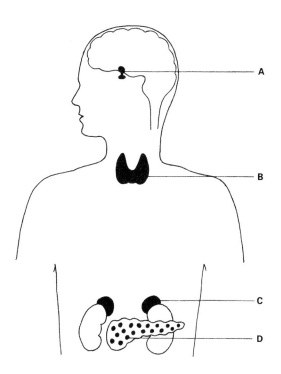

73

6 Which is the site of glucagon production?

7 Which is the site of anti-diuretic hormone production?

8 Which of these glands makes adrenaline?

Items **9** and **10** refer to the following graph which shows the blood sugar level of a person who has consumed 50 g of glucose at the time indicated.

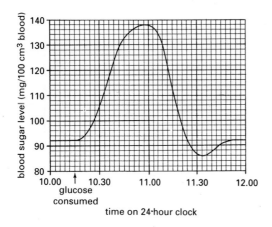

9 The maximum level of blood sugar, expressed as a multiple of the normal (resting) value, is
A 0.5.   B 0.67.   C 1.5.   D 12 696.

10 Compared with normal levels, extra insulin and extra glucagon would be present at certain times during this two-hour period. Two of these times are

	extra insulin	extra glucagon
A	10.30	11.00
B	10.30	11.30
C	11.00	10.30
D	11.30	10.30

11 The temperature-monitoring centre in the human brain is situated in the
A cerebellum.   B pituitary.
C medulla.   D hypothalamus.

12 Following overheating of a mammal's internal environment, the skin acts as an effector. Which of the following changes occur under these circumstances?

	altered state of arterioles leading to skin	altered state of erector muscles in skin
A	constricted	relaxed
B	dilated	relaxed
C	dilated	contracted
D	constricted	contracted

Items **13**, **14** and **15** refer to the diagram below.

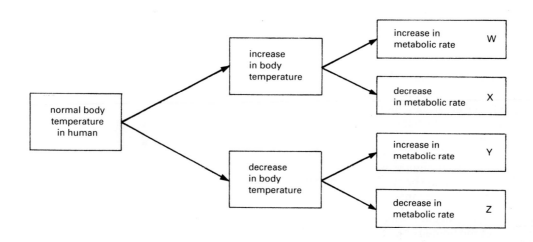

13 Which letters indicate the normal negative feedback control of body temperature?
   A W and Y   B W and Z
   C X and Y   D X and Z

14 Which situation would be the immediate result of exposure to intense cold?
   A W   B X   C Y   D Z

15 Which situation would be the result of prolonged exposure to intense cold leading to hypothermia?
   A W   B X   C Y   D Z

Items 16 and 17 refer to the following possible answers.
   A whale and herring   B herring and shark
   C shark and dolphin   D dolphin and whale

16 Which of these animals are *both* endotherms?

17 Which of these animals are *both* ectotherms?

18 Bacteria taken from the intestine of a certain animal were found to grow on a culture plate at 7°C, but not at 37°C. From which animal had they come?
   A cod   B seal
   C penguin   D polar bear

19 Which of the following graphs best shows the relationship between external temperature and body temperature for an endotherm and an ectotherm?

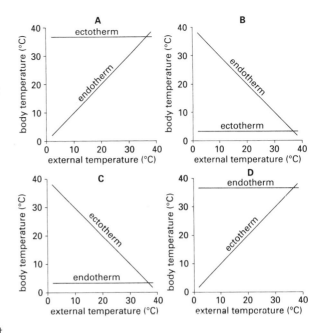

20 The responses to external environment shown in the following table are typical of one of the animals listed below the table. Which one?

temperature of habitat (°C)	activity of animal	metabolic rate
0	inactive	low
10	inactive	low
20	active	normal
30	very active	high

   A gerbil   B lizard
   C vulture   D camel

Test 32 Population dynamics – part 1

Items **1**, **2** and **3** refer to the following possible answers.
A dynamic equilibrium
B carrying capacity
C population dynamics
D environmental resistance

1 Which term means the upper limit in size of a population that can be maintained by the resources available in its ecosystem?

2 Which expression refers collectively to those factors that prevent a population from increasing in size indefinitely?

3 Which term means the study of population changes and the factors that cause such changes?

Items **4** and **5** refer to the following graph which plots the birth rates and death rates for a certain country over a period of 80 years.

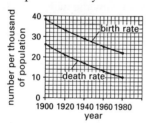

4 What was the percentage annual increase in population during this period of time?
A 0.6   B 1.2   C 6   D 12

5 After 1980 the population continued to increase in size but showed a decrease in rate of growth. Which of the following graphs best represents this trend?

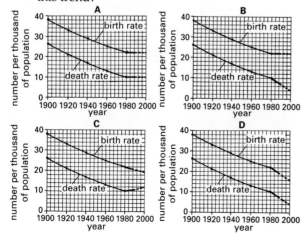

6 Which of the following factors acts on a population in a density-dependent manner?
A drought   B pollution
C flooding   D predation

7 Which of the following acts as a density-independent factor on a population?
A forest fire   B shortage of food
C disease   D excretory wastes

8 The data in the following table refer to the results from a series of experiments using water fleas.

population density of adults (number/ litre of pond water)	average number of offspring produced per day by female aged:		
	10 days	15 days	20 days
1 000	2.5	3.2	3.7
4 000	1.6	1.9	2.1
8 000	0.7	1.1	1.3
16 000	0.4	0.5	0.7

From this data it can be concluded that
A birth rate falls with increased crowding which acts in a density-dependent manner.
B birth rate rises with increased crowding which acts in a density-dependent manner.
C birth rate falls with increased crowding which acts in a density-independent manner.
D birth rate rises with increased crowding which acts in a density-independent manner.

9 Which of the following graphs best represents a typical predator-prey relationship?

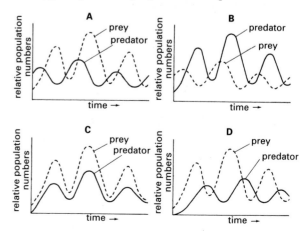

10 Which of the following diagrams correctly represents homeostatic control of population size?

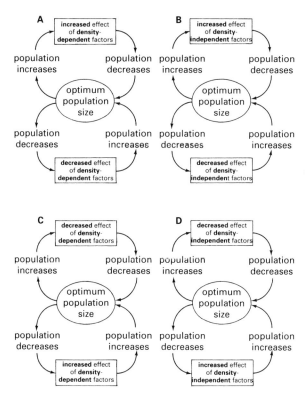

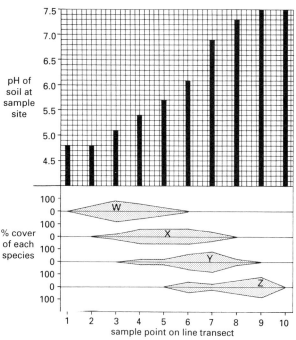

Items **11**, **12** and **13** refer to the diagram at the top of the next column, which charts the results from an experiment to investigate the effect of soil pH on the distribution of four plant species (W, X, Y and Z).

The soil pH and percentage cover of each plant species were recorded at each of ten sample points at regular intervals along a line transect. The plant cover data are presented as kite diagrams (i.e. symmetrical line graphs on either side of a base line).

11 Only two of the plant species were recorded at sample point
    A 3.   **B** 4.   **C** 5.   **D** 6.

12 Species W, X and Y were all found to be growing in soil of pH
    A 5.1   **B** 5.7   **C** 6.1   **D** 6.9

13 Which species is able to tolerate the widest range of soil pH?
    A W   **B** X   **C** Y   **D** Z

Items **14** and **15** refer to the following information.

A biologist wished to estimate the total population number of a species of moth living in a wood. She captured 40 moths, marked them with non-toxic paint and released them. The next night she set up a light trap and caught 100 specimens of the moth. Of these, 8 bore the paint mark.

14 Based on these figures, the total population number of this species of moth is
    A 200.   **B** 320.   **C** 500.   **D** 4000.

15 If m^1 = number caught and marked on the first night,
      y = total captured on the second night,
      m^2 = number captured on the second night bearing the paint mark
and x = total number in the population,
then a simple formula for the estimation of x is

A $\dfrac{x}{m^2} = \dfrac{y}{m^1}$ **B** $\dfrac{m^1}{x} = \dfrac{y}{m^2}$

C $\dfrac{m^2}{x} = \dfrac{m^1}{y}$ **D** $\dfrac{x}{m^1} = \dfrac{y}{m^2}$

Test 33 Population dynamics – part 2

1. An indicator organism is one which
   A is always restricted to life in badly polluted environments.
   B provides information about the maximum sustainable yield of a food species.
   C shows that the members of a certain species are threatened with extinction.
   D can be used to quantify relative levels of pollution in the biosphere.

Items **2** and **3** refer to the following graph which charts the effect of increased intensity of fishing on four species of edible fish caught in the North Sea. (Sustained yield means the maximum yield of fish that can be maintained from year to year.)

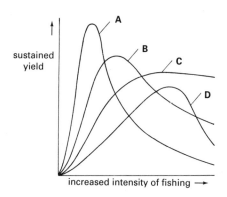

2. Which species is *least* affected by increased intensity of fishing?

3. Which species of fish is most likely to be the slowest-growing one?

Items **4, 5** and **6** refer to the following passage.

The spiderwort is a common wild flowering plant found growing in many parts of North America. It has long been known to act as an indicator of environmental degradation
5 caused by excesses of sulphur dioxide, motor engine exhaust fumes and pesticides. However, recently it has also revealed itself to be an ultra-sensitive monitor of ionising radiation.
10 Human normally receive about 100 millirems of background radiation per year. When spiderwort is exposed to as low a level as 150 millirems, the colour of the hair cells on its stamens changes from blue to pink. This
15 occurs as a result of the destruction by radiation of the genetic material needed to produce blue pigment. Whereas most organisms normally take a long time to show any noticeable effects in
20 response to low (but unacceptable) levels of radiation, spiderwort is so sensitive that its colour change occurs within 10–18 days of exposure.

4. Which of the following is the most appropriate title for the passage?
   A Useful indicator organisms
   B Floral monitor of radiation
   C Forms of atmospheric pollution
   D Mutation frequency in spiderwort

5. Which of the following is *not* referred to in the passage?
   A radioactive fallout
   B atmospheric pollution
   C excessive use of sprays in agriculture
   D effect of organic waste on rivers

6. Which numbered lines in the passage contain information that verifies that radiation acts as a mutagenic agent?
   A 8–9   B 10–11   C 15–16   D 19–20

Items **7** and **8** refer to the following table which shows the concentration of a non-biodegradable pesticide residue in the tissues of several organisms in a food chain, and in the water of their ecosystem.

	concentration of pesticide (ppm)
water	0.00005
plankton	0.04
herbivorous fish	0.23
carnivorous fish	2.07
fish-eating bird	6.00

7. The concentration of pesticide increased by a factor of 9 between
   A herbivorous fish and carnivorous fish.
   B carnivorous fish and fish-eating bird.
   C plankton and herbivorous fish.
   D water and plankton.

8. The concentration of pesticide in the fish-eating bird is greater than that in the water by a factor of
   A 1.2×10^3.   B 1.2×10^4.
   C 1.2×10^5.   D 1.2×10^6.

Items **9** and **10** refer to the following graph which shows the results from a survey done on the number of lichen species growing along a 20 km transect, from the centre of a city out to a country area.

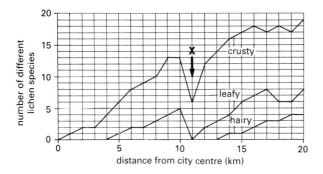

9 The dip in the graph at arrow X indicates
  A an area of especially clean air.
  B a local increase in sulphur dioxide concentration of air.
  C an area lacking both hairy and crusty lichens.
  D a lower level of atmospheric pollution compared with the country area.

10 Twenty-eight different species of lichen were recorded at one of the sites. The distance (in km) of this site from the city centre was
  A 16.   B 17.   C 18.   D 19.

Test 34 Population dynamics – part 3

1 The following statements refer to the process of ecological succession. Which one is *incorrect*?
  A A new group of plant species achieves dominance and ousts the old ones.
  B The height and biomass of the vegetation increases as the process proceeds.
  C Each group of species modifies the habitat making it more favourable for some other species.
  D The number and variety of animal species decreases as the process proceeds.

2 Primary succession would occur on
  A an abandoned field.
  B a disused railway line.
  C a rock surface bared by glaciation.
  D a garden neglected by its owner.

3 Which of the following is *not* a characteristic of a climax community?
  A It is the final product of long term change within the community.
  B It is an immature community still to reach dynamic equilibrium with the environment.
  C It is self-perpetuating and not replaced by another community.
  D It is a stable community which does not show directional change.

4 As a newly-bared mountain scree underwent succession, it became populated by the plant communities listed below.
  1 shrubs
  2 mosses
  3 small trees
  4 lichens
  5 grasses
  These communities would have appeared in the order
  A 2, 4, 5, 1, 3.   B 2, 5, 4, 3, 1.
  C 4, 2, 5, 1, 3.   D 4, 5, 2, 3, 1.

5 The natural climatic climax community in lowland Scotland is
  A forest.   B grassland.
  C moorland.   D peat bog.

6 The natural climax community typical of a geographical region is determined by *both*
  A prevailing climate and soil type.
  B soil type and land reclamation.
  C land reclamation and agricultural practices.
  D agricultural practices and prevailing climate.

7 Which of the following is *not* a climax community?
  A tundra   B tropical rainforest
  C heath moorland   D field of grass

8 In most parts of Britain, the vegetation is prevented from reverting to the natural climax community by
   A human activities.
   B prevailing climate.
   C underlying edaphic factors.
   D competition between species.

9 The sequence in which this series of events would occur is
   A R, S, U, Q, P, T.   B R, S, Q, U, P, T.
   C R, Q, S, U, P, T.   D R, S, Q, P, U, T.

10 Which of the following statements is *not* correct?
   A Tussock-forming sedges make the environment more favourable for alder trees.
   B Aquatic plants make the environment more favourable for reeds.
   C Alder trees make the environment more favourable for oak trees.
   D Sedges make the environment more favourable for reeds.

Items 9 and 10 refer to the following information.
   A fresh water lake may undergo plant succession if it has an inflow stream which carries silt as shown in the diagrams below.

alder trees grow in marshy soil and increase its nitrogen content

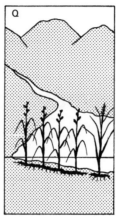

sedges thrive on accumulated silt

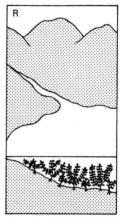

aquatic plants interfere with free flow of water making it deposit silt

reeds with underground rhizomes grow in silt and gather more sediment

oak forest becomes established on fertile soil

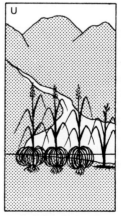

tussock-forming sedges form mounds of marshy soil made of silt and humus

Test 35 Maintaining a water balance – animals

Items **1** and **2** refer to the following possible answers.
A jellyfish **B** haddock
C *Paramecium* **D** stickleback

1 Which animal possesses body contents which are hypotonic to its natural surroundings?

2 Which animal's body contents are isotonic to its natural surroundings?

3 Which of the osmoregulatory mechanisms shown in the diagram below is employed by a fresh-water bony fish?

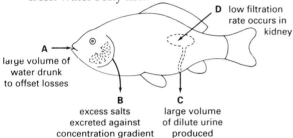

4 Which of the following pairs of characteristics would be true of a mammal which produces very concentrated urine?

	level of ADH in blood	relative length of loops of Henle
A	low	short
B	low	long
C	high	short
D	high	long

Items **5** and **6** refer to the possible answers in the following table

	relative amount of water drunk	relative amount of urine produced
A	none	little
B	little	much
C	much	much
D	much	little

5 Which animal is a tuna fish?

6 Which animal is a desert rat?

7 The gills of a salt water fish
  A lose water by osmosis, and absorb salts.
  B gain water by osmosis, and absorb salts.
  C lose water by osmosis, and excrete salts.
  D gain water by osmosis, and excrete salts.

8 The structures shown in the following diagram are present in the kidney of a certain animal.

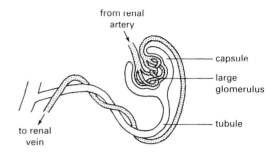

This animal could be a
A camel.
B desert rat.
C fresh water bony fish.
D salt water bony fish.

Items **9** and **10** refer to the following graph which charts the variation in salt concentration of the body fluid of four invertebrate animals when placed in different dilutions of sea water.

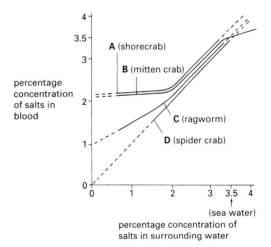

9 Which animal is completely unable to maintain a blood concentration of salt above that of the surrounding water?

10 Which animal is able to maintain a blood concentration of salt lower than that of the surrounding water?

11 The four routes by which water is lost from a horse are shown in the diagram below. Which of these does *not* occur in a kangaroo rat?

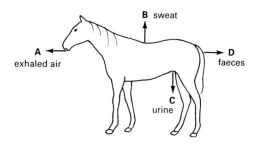

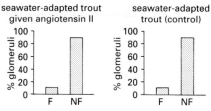

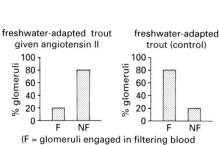

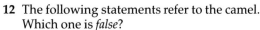

(F = glomeruli engaged in filtering blood
NF = non filtering glomeruli)

12 The following statements refer to the camel. Which one is *false*?
A Its red blood cells are very sensitive to osmotic changes of the blood.
B It can drink 100 litres of water in a very short space of time.
C Its rate of urine production is low and little water is lost in faeces.
D It does not begin to sweat until its body temperature reaches 41°C.

13 Which of the following does *not* occur when an Atlantic salmon migrates from fresh to salt water?
A change in filtration rate of kidneys
B extension in length of kidney tubules
C reversal in direction of salt transfer by gills
D reduction in volume of urine produced

Items **14** and **15** refer to the following graphs which summarise the results from a series of experiments on seawater-adapted and freshwater-adapted trout treated with a hormone called angiotensin II.

14 From this experiment it can be concluded that an increase in concentration of angiotensin II results in
A a reduction in number of filtering glomeruli in seawater-adapted trout.
B an increase in number of filtering glomeruli in seawater-adapted trout.
C a reduction in number of filtering glomeruli in freshwater-adapted trout.
D an increase in number of filtering glomeruli in freshwater-adapted trout.

15 In some fish, angiotensin II induces an antidiuretic effect, thereby promoting water retention. This effect could be brought about by
A vasoconstriction of renal arterioles which reduces rate of blood flow to glomeruli.
B vasodilation of renal arterioles which reduces rate of blood flow to glomeruli.
C vasoconstriction of renal arterioles which increases rate of blood flow to glomeruli.
D vasodilation of renal arterioles which increases rate of blood flow to glomeruli.

Test 36 Maintaining a water balance – plants

1. The ascent of sap in a plant is thought to be assisted by
   1 transpiration pull,
   2 root pressure,
   3 capillarity.
   Which of the above involve osmosis?
   A 1 and 2 only **B** 1 and 3 only
   C 2 and 3 only **D** 1, 2 and 3

2. The column of sap being pulled up a tree is maintained without a break because the water molecules
   A adhere to one another.
   B cohere to one another.
   C adhere to the surrounding air molecules.
   D cohere to the sides of the xylem vessels.

Items **3** and **4** refer to the following diagram of part of a transverse section of a leaf.

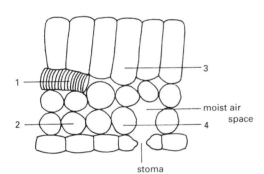

3. At which location is the water concentration highest?
   A 1 **B** 2 **C** 3 **D** 4

4. For the transpiration pull to be effective, a water concentration gradient must exist between
   A 1 and 3.   **B** 1 and 4.
   C 2 and 3.   **D** 2 and 4.

5. During a period of very rapid transpiration, the diameter of a tree trunk often
   A increases because less water is present in the xylem vessels.
   B decreases because more water is present in the xylem vessels.
   C increases because tension decreases in the xylem vessels.
   D decreases because tension increases in the xylem vessels.

Items **6** and **7** refer to the following graph.

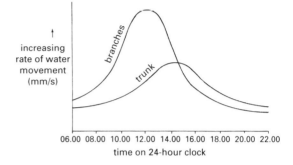

6. From the graph it can be correctly concluded that during the course of one day
   A the increase in rate of water movement first begins in the trunk.
   B the greatest increase in rate of water flow first occurs in the trunk.
   C any change that first occurs in the trunk is repeated later in the branches.
   D the greatest increase in rate of water flow first occurs in the branches.

7. The results presented in this graph support the theory that
   A the greatest rate of transpiration normally occurs at 14.00 hours.
   B the rate of water movement in the branches is not related to that in the trunk.
   C the leaves provide the driving force for the upward movement of water.
   D a larger amount of water passes through the branches than the trunk.

8. A stomatal pore opens as a result of the guard cells
   A absorbing water by osmosis.
   B using up sugar during respiration.
   C using up water during photosynthesis.
   D absorbing carbon dioxide by diffusion.

9. The following graph shows the rates of water absorption and transpiration that occurred in a sunflower plant during a 24-hour period.

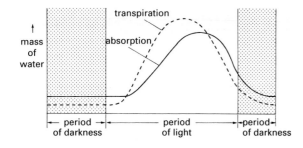

From this evidence it is true to say that in
A light, absorption rate always exceeds transpiration rate.
B darkness, absorption rate always exceeds transpiration rate.
C light, transpiration rate always exceeds absorption rate.
D darkness, transpiration rate always exceeds absorption rate.

10 The following diagram shows four different pieces of apparatus.

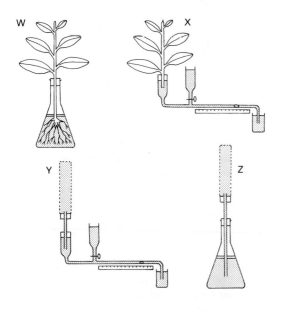

Which of these are *both* atmometers?
A W and X
B X and Y
C W and Z
D Y and Z

11 Which of the following is *not* an essential precaution taken when setting up and using a bubble potometer?
A cutting the plant shoot under water
B ensuring that the entire system is kept airtight
C using a plant that bears both leaves and flowers
D preventing the air bubble from reaching and entering the shoot

12 The bubble in a bubble potometer moves most rapidly when the apparatus is in conditions which are
A dark and still.   B dark and windy.
C light and still.   D light and windy.

13 A weight atmometer loses *least* weight when exposed to conditions which are
A cool and damp.   B cool and dry.
C hot and damp.   D hot and dry.

14 In which of the following do *both* factors affect the rate of water uptake in a potometer but *not* in an atmometer?
A relative humidity and air movement
B light intensity and stomatal closure
C relative humidity and light intensity
D air movement and stomatal closure

15 The following graph charts the rate of transpiration from a geranium plant's leaves. When did the plant's stomata begin to open?

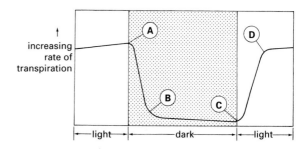

16 The table below gives the rates of transpiration for three different plants.

	rate of transpiration (mg water/cm² leaf/hr)	
	light	dark
holly	20	0
heather	10	0
sycamore	30	1

Which of the following is *not* a reasonable explanation of these results?

A Some stomata remain partly open in darkness.
B Cuticular transpiration occurs from some leaves.
C Transpiration always comes to a halt in darkness.
D A little water escapes from a leaf's thin edges.

Items **17** and **18** refer to the following diagram of part of a leaf from a xerophyte.

Items **19** and **20** refer to the following diagram of the pondweed *Potamogeton*.

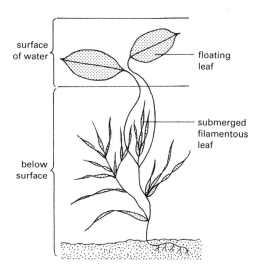

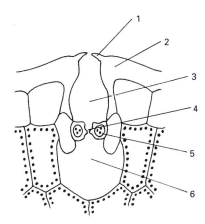

17 The sunken stoma is labelled
    A 3.   B 4.   C 5.   D 6.

18 Which of the following is *not* a xeromorphic feature?
    A 1   B 2   C 3   D 6

19 Which of the descriptions in the following table refers to a floating leaf?

	number of stomata on upper surface	number of stomata on lower surface
A	many	none
B	none	many
C	none	none
D	many	many

20 Which of the following statements is *false*? The submerged filamentous leaves
    A offer little resistance to rapid water flow.
    B present a large surface area for light absorption.
    C possess many xylem vessels for support.
    D take in mineral salts over a big surface area.

Test 37 Obtaining food – animals

1 Which of the following diagrams *best* represents the trail followed by an ant when foraging for food and then, on finding it, returning to the colony?

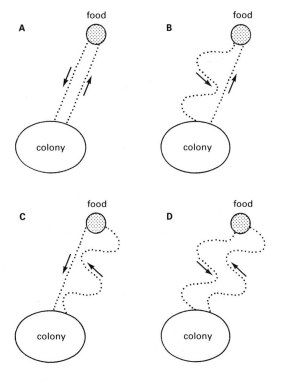

The diagram below shows a five and six food supplies.

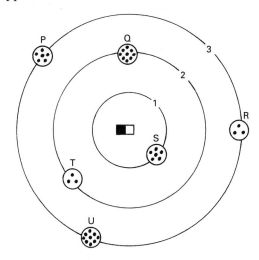

KEY

☐ = hive

■ = shadow of hive

1-3 = contours at 100 m intervals from hive

(··) = 30 flowers ripe with pollen and nectar

(:··) = 60 flowers ripe with pollen and nectar

(:··:) = 90 flowers ripe with pollen and nectar

Items **2**, **3** and **4** refer to the following information and diagram.

On returning to the hive from a successful foraging trip, a bee does a waggle dance which communicates vital information to other bees as follows.

feature of dance	information conveyed
number of turns/min	distance of food from hive (the faster the dance, the closer the food)
indication of angle between sun and food supply	direction of food from hive
duration of dance	richness of food supply (the longer the dance, the richer the food)

2 Which of the following is the richest source of food located furthest from the hive?
  A P B Q C R D U

3 Which of the following angles (measured in a clockwise direction) would direct bees to food source P?
  A 50° B 90° C 140° D 220°

4 Which of the following bees is indicating a food source *not* shown in the diagram?

bee	speed of dance	angle (measured in clockwise direction)	duration of dance
A	fast	240°	long
B	medium	140°	short
C	fast	40°	medium
D	slow	0°	short

86

5 Which of the following tabulated descriptions of foraging behaviour *best* applies to a pride of lionesses hunting in a savannah ecosystem rich in game?

	search time	pursuit time	choice of prey
A	short	long	selective
B	short	long	unselective
C	long	short	selective
D	long	short	unselective

6 Which of the following is *not* a feature of foraging behaviour?
  A Energy can be saved by searching for food in a group.
  B Energy supply can be maintained by defending a territory.
  C Energy derived from prey must be equal to that expended in the search.
  D Energy sources of low quality are often ignored in favour of high quality prey.

7 The following graph shows the results from an investigation into the relationship between prey length and energy gained by a certain species of predatory bird.

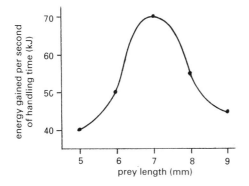

These results illustrate the principle that
  A the smallest prey is the most profitable in terms of net energy gain, since it is the easiest to catch and kill.
  B the medium-sized prey provides the largest net energy gain, since it contains most energy relative to the energy expended on the hunt.
  C the largest prey contains most energy and offers the predator the greatest net energy gain for the effort required to subdue it.
  D all sizes of prey provide similar net energy gains if an appropriate number are caught and consumed by the predator.

8 Competition between members of different species is called
  A interspecific, and is normally less intense than intraspecific competition.
  B interspecific, and is normally more intense than intraspecific competition.
  C intraspecific, and is normally less intense than interspecific competition.
  D intraspecific, and is normally more intense than interspecific competition.

9 Which of the following graphs illustrates the result of competition between a successful species and an unsuccessful species of animal for the same source of food?

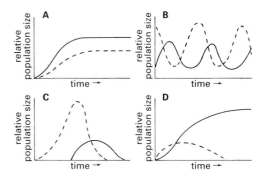

10 The diagrams below show the same area of moorland containing territories inhabited by red grouse in different years. Which diagram *best* represents the year in which the food supply was most plentiful?

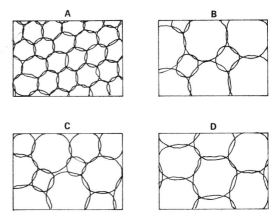

11 The graph below shows the increase in a population of fieldmice with time.

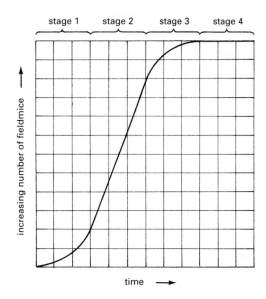

Which of the following explanations *fails* to account for the change occurring at that particular stage of the graph?

	stage	possible explanation
A	1	predators have been removed
B	2	food supply has become limiting
C	3	interspecific competition has increased
D	4	overcrowding has caused disease to spread

12 The following graph refers to the economics of defence of different-sized territories for a certain species of animal.

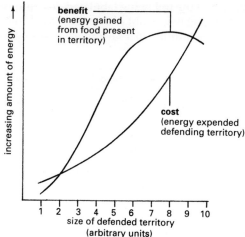

The optimum size of territory for this species, under these conditions of food availability, is
A 4.   B 5.   C 6.   D 7.

13 The following list refers to pecking behaviour observed amongst six hens (P, Q, R, S, T and U).

P pecked U   S pecked U
P pecked T   T pecked Q
R pecked T   U pecked R
S pecked P   U pecked Q

Which bird was *third* in the peck order?
A P   B U   C R   D T

14 Which of the following does *not* occur as a direct result of a herd of animals being dominated by one older male?
A minimum aggression amongst the group members
B experienced leadership during times of crisis
C promotion of the group's chance of survival
D equal choice of food for all herd members

15 Which of the following are *both* subordinate responses shown by a young wolf in the presence of the pack leader?
A head lowered and ears cocked
B hackles raised and tail lowered
C ears flattened and eyes averted
D eyes staring and teeth bared

Test 38 Obtaining food – plants

Items **1** and **2** refer to the following possible answers.
A sessile and autotrophic
B sessile and heterotrophic
C mobile and autotrophic
D mobile and heterotrophic

1. Which description applies to the vast majority of flowering plants?

2. Which description is true of all advanced animals?

3. The leaves of many sun plants grow in such a way that each leaf blade is held at a certain angle to the rays of the mid-day sun.
   Which of the following would be the most effective angle for light absorption?

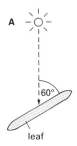
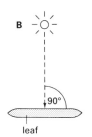
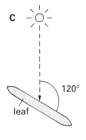
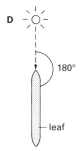

4. Competition between members of the same species is called
   A intraspecific, and is more intense than interspecific competition.
   B intraspecific, and is less intense than interspecific competition.
   C interspecific, and is more intense than intraspecific competition.
   D interspecific, and is less intense than intraspecific competition.

5. Which of the following leaf mosaic patterns would be the most effective?

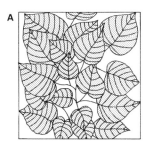
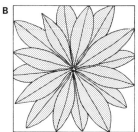
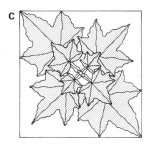
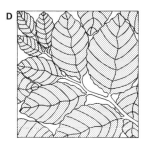

6. *Galium saxatile* grows best in acidic soil but will grow on alkaline soil in the absence of competition. *Galium pumilim*, on the other hand, grows best on alkaline soil but can survive on acidic soil.
   In a competition experiment, equal numbers of seeds of both species of *Galium* were planted together in two pots. Pot 1 contained alkaline soil and pot 2 contained acidic soil.
   Which of the following sets of results is the most likely outcome of this experiment?

	pot 1		pot 2	
	G. saxatile	G. pumilum	G. saxatile	G. pumilum
A	+	−	−	+
B	−	+	+	−
C	+	−	+	−
D	−	+	−	+

(+ = *growth*; − = *no growth*)

7 The graph below refers to the net amount of sugar produced or used in a green plant leaf over a 24-hour period.

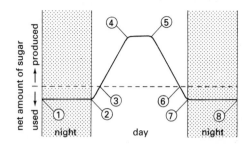

Which of the following are *both* compensation points?
A 1 and 8   B 2 and 7
C 3 and 6   D 4 and 5

8 The following list gives four situations that could affect a mixed community of plants growing in a grassland ecosystem.
1 herbivores graze unselectively on all species
2 all herbivores are removed from the ecosystem
3 herbivores graze selectively on delicate plant varieties
4 herbivores graze selectively on sturdy dominant grasses.

Which two situations would tend to maintain species diversity amongst the plant community?
A 1 and 3   B 2 and 3
C 1 and 4   D 2 and 4

Items **9** and **10** refer to the graph at the bottom of the page.

9 Which of the following statements is *correct*?
A X is the sun plant and its compensation point occurred at 3 units of light.
B X is the shade plant and its compensation point occurred at 5 units of light.
C Y is the sun plant and its compensation point occurred at 5 units of light.
D Y is the shade plant and its compensation point occurred at 3 units of light.

10 The length of plant X's compensation period in hours was
A 1.5   B 2.5   C 4.0   D 5.0

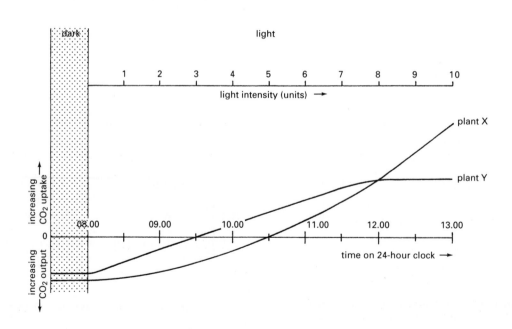

Test 39 Coping with dangers – animals

Items **1** and **2** refer to the following information.
When a tube worm is touched, it responds by withdrawing into its tube. The graph below charts the results from an experiment where 60 tube worms were subjected to repeated touching over 20 trials.

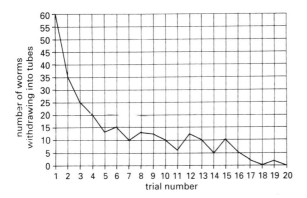

1 At which trial number did 25% of the worms show the escape response?
 A 3 **B** 6 **C** 10 **D** 15

2 At which trial number had 35 of the worms become habituated?
 A 2 **B** 3 **C** 4 **D** 5

3 Habituation is a beneficial form of behaviour because it enables an animal to
 A avoid a potentially dangerous situation.
 B defend itself against an enemy.
 C effect its escape response quickly.
 D conserve energy for essential activities.

4 It is essential that habituation is a short-lived form of learning, otherwise the animal could
 A be left open to danger.
 B expend energy needlessly.
 C exhaust its food reserves.
 D forget how to effect its escape response.

5 Young children are taught by their parents not to poke their fingers into electrical sockets. To be of protective value, it is essential that such learning involves a
 A short term modification of the response made to the stimulus.
 B long term modification of the response made to the stimulus.
 C short term modification of the stimulus made to the response.
 D long term modification of the stimulus made to the response.

6 Which of the following graphs correctly represents a learning curve?

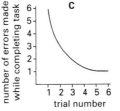 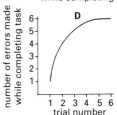

7 Which of the following is a passive form of defence?
 A performing a 'broken wing' distraction display
 B rolling over and feigning death
 C fleeing from the enemy at top speed
 D possessing a body resembling nearby plants

Items **8** and **9** refer to the following possible answers.
A mimicry
B disruptive colouring
C counter shading
D menacing eye markings

8 Which of these is possessed by a trout?

9 Which of these is employed by non-poisonous striped insects?

10 Which of the following is a social mechanism for defence?
 A alarm calls in birds
 B spines on bodies of hedgehogs
 C poison glands in snakes
 D foul-smelling secretions squirted by skunks

Test 40 Coping with danger – plants

Items **1**, **2** and **3** refer to the graph below. It charts the amounts of two types of chemical present in bracken leaf fronds which begin to push up out of the soil in spring.

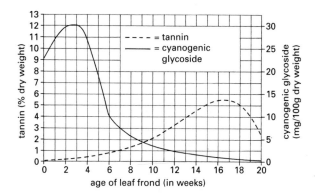

1 Which of the following conclusions can be correctly drawn from the graph?
  A Younger leaves possess a high concentration of cyanogenic glycoside and a low concentration of tannin.
  B Older leaves possess a high concentration of cyanogenic glycoside and a low concentration of tannin.
  C Younger leaves possess a high concentration of both cyanogenic glycoside and tannin.
  D Older leaves possess a high concentration of both cyanogenic glycoside and tannin.

2 The amounts of the two chemicals present in leaves at week 7 are

	tannin (% dry weight)	cyanogenic glycoside (mg/100 g dry weight)
A	1.0	3.0
B	1.0	7.5
C	3.0	2.5
D	7.5	1.0

3 The concentration of cyanogenic glycoside shows the biggest drop in concentration between weeks
  A 3 and 4.   **B** 4 and 5.
  C 5 and 6.   **D** 6 and 7.

4 Which of the following equations correctly represents the process of cyanogenesis?
  A toxic cyanogenic glycoside $\xrightarrow{\text{enzyme action}}$ toxic hydrogen cyanide
  B non toxic cyanogenic glycoside $\xrightarrow{\text{enzyme action}}$ toxic hydrogen cyanide
  C toxic hydrogen cyanide $\xrightarrow{\text{enzyme action}}$ toxic cyanogenic glycoside
  D non toxic hydrogen cyanide $\xrightarrow{\text{enzyme action}}$ toxic cyanogenic glycoside

5 Which of the following is a modified side branch?
  A hair   **B** spine   **C** sting   **D** thorn

6 The diagram below illustrates a desert plant called *Echinocactus*.

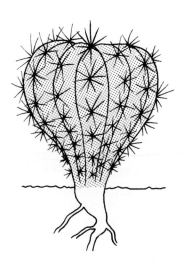

This cactus is protected from thirsty mammals by possessing
  A a rounded shape.
  B a thick waxy cuticle.
  C leaves reduced to spines.
  D water stored in succulent tissues.

Items **7** and **8** refer to the labelled structures in the following diagram of four different types of plant.

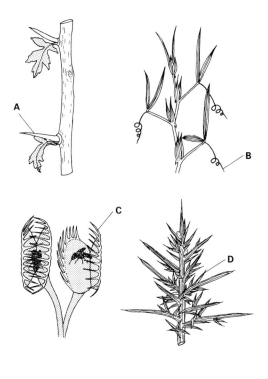

7 Which structure is a protective spine?

8 Which structure is a thorn?

9 The diagram below shows four common weeds (not drawn to scale). Which would be *least* likely to survive on a meadow heavily grazed by cattle?

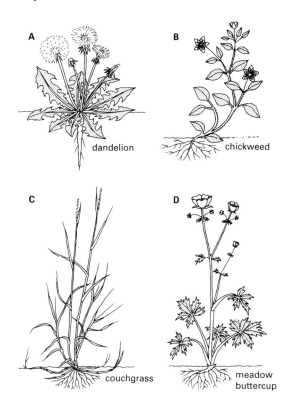

10 Which of the following graphs *best* represents the relationship between number of spines per holly leaf and height of leaf from ground?

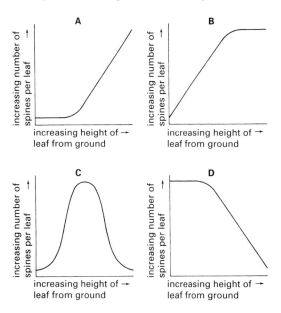

Specimen Examination Paper

1 The graph below shows the relationship between ion uptake by dandelion root cells and distance from root apex.

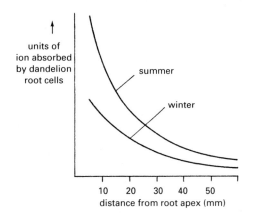

From these results it can be correctly concluded that in dandelion roots, most ion uptake occurs in
A young newly formed cells in summer.
B young newly formed cells in winter.
C older differentiated cells in summer.
D older differentiated cells in winter.

2 Which of the following acids is the final product of glycolysis?
A citric   B lactic
C pyruvic   D oxaloacetic

3 Which of the cells below would possess the greatest number of mitochondria per unit volume of cell?
A guard cell
B motile sperm
C red blood cell
D palisade mesophyll

4 An area of grassland under study by scientists was found to receive 2×10^6 kJ of solar energy over a period of one year. The fate of the energy *not* used for photosynthesis was found to be as follows.

reflected by leaves	2.2×10^5 kJ
transmitted to the ground	8.0×10^4 kJ
converted to heat and lost	1.6×10^6 kJ

The percentage of light used annually for photosynthesis was
A 1.   B 2.   C 5.   D 10.

5 Which of the following refers to photolysis in a green plant cell?

	process	location in chloroplast
A	generation of ATP	stroma
B	fixation of carbon	grana
C	evolution of oxygen	stroma
D	reduction of hydrogen acceptor	grana

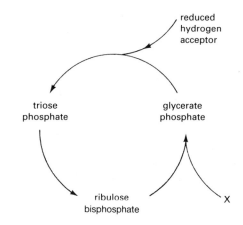

6 The diagram above shows a simplified version of the Calvin cycle.
The correct identity of substance X is
A carbon dioxide.   B glucose.
C oxygen.   D water.

7 If a DNA molecule contains 5000 base molecules of which 22% are adenine, then the number of guanine molecules present is
A 1200.   B 1400.   C 2400.   D 2800.

8 Successful replication of chromosomes does *not* require the presence of
A ribosomes.
B a DNA template.
C nuclear enzymes.
D adenosine triphosphate.

9 The following list gives four stages involved in protein synthesis.
1 tRNA and mRNA meet in a ribosome.
2 Region of DNA opens up exposing bases.
3 mRNA is translated into a polypeptide chain.
4 mRNA is transcribed using free nucleotides.
These would occur in the sequence:
A 2, 1, 4, 3.   B 2, 4, 1, 3.
C 4, 2, 1, 3.   D 4, 1, 3, 2.

10 All viruses contain
   A DNA.
   B RNA.
   C DNA and RNA.
   D DNA or RNA.

11 Which of the following statements is correct?
   A A B-lymphocyte produces cell-bound antibodies which are active against several different antigens.
   B A T-lymphocyte produces free antibodies which are active against one specific antigen.
   C A B-lymphocyte produces free antibodies which are active against one specific antigen.
   D A T-lymphocyte produces cell-bound antibodies which are active against several different antigens.

12 Which of the following does *not* occur during the first meiotic division?
   A complete separation of chromatids
   B crossing over between chromatids
   C pairing of homologous chromosomes
   D reduction of chromosome number

13 Syndactyly is a genetically inherited condition in humans involving the joining of two or more fingers by a web of skin and muscle. It is determined by the presence of a dominant allele.

   The key to the symbols used in the family tree shown below is as follows:

   ■ = syndactylous male
   ● = syndactylous female
   □ = normal male
   ○ = normal female

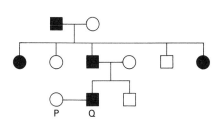

   The chance of individuals P and Q producing a syndactylous child is
   A 1 in 1.   B 1 in 2.   C 1 in 3.   D 1 in 4.

14 In a certain type of guinea pig, long hair (L) is dominant to short hair (l), and straight hair (S) is dominant to wavy hair (s). It is found that the cross LlSs × llss results in the ratio 1 long, straight : 1 long, wavy : 1 short, straight : 1 short, wavy.

   Which of the following crosses would also produce this result?
   A LLSs × llss   B Llss × llSS
   C Liss × llSs   D LlSs × LlSs

15 Red-green colour blindness is a sex-linked trait in humans. A colour blind man marries a woman who is heterozygous for this condition.
   What proportion of their sons are likely to be colour blind?
   A 25%   B 50%   C 75%   D 100%

16 Three true-breeding female fruit flies with red eyes and black bodies were crossed with six true-breeding male pink-eyed, grey-bodied flies. All members of the F_1 generation were found to possess red eyes and grey bodies.

   Three F_1 females were crossed with six F_1 males and produced an F_2 generation consisting of 115 red-eyed, grey : 37 pink-eyed, grey : 41 red-eyed, black : 12 pink-eyed, black.

   The results indicate that inheritance of these two characteristics is determined by
   A sex-linked genes.
   B linked autosomal genes.
   C genes on independently segregating chromosomes.
   D one gene affecting both characteristics and showing incomplete dominance.

17 The following diagram shows two types of chromosome mutation.

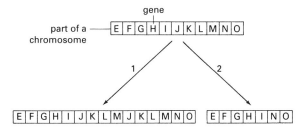

   These are called

	1	2
A	duplication	deletion
B	duplication	substitution
C	inversion	deletion
D	inversion	substitution

18 When first brought into clinical use, the antibiotic penicillin was found to be highly

effective against the bacterium *Staphylococcus*, which can cause many serious infections. However, within ten years, many strains of *Staphylococcus* were found to be resistant to penicillin.

This situation arose because

A the bacterium's original gene pool already contained some alleles which determined resistance to penicillin.
B penicillin induced a mutation producing a resistance gene which was passed on to future generations of bacteria.
C all bacterial cells possess a few genes which are resistant to penicillin and enjoy a selective advantage.
D penicillin stimulated the development of resistance in some bacteria and these were later selected during infection.

19 The evolution of various breeds of cattle is shown in the following diagram.

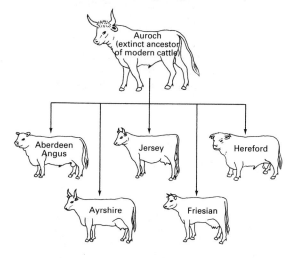

This has occurred mainly by

A natural selection of existing alleles in the gene pool.
B natural selection of newly mutated genes.
C artificial selection of existing alleles in the gene pool.
D artificial selection of newly mutated alleles.

20 Continuous inbreeding in an isolated population leads to

A increase in phenotypic variation.
B decrease in population size.
C increase in mutation rate.
D decrease in genetic variation.

21 The following diagram shows a piece of human DNA ready to be sealed into a bacterial plasmid.

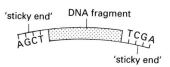

Which of the diagrams below represents a suitable plasmid?

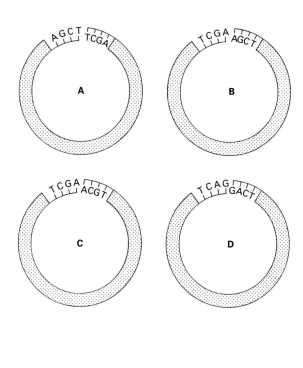

22 Which of the following is *not* a source of heritable variation?

A independent assortment of chromosomes
B crossing-over during meiosis
C gene mutations
D natural selection

23 In which of the following pairs are the structures homologous?

A eyes of beetles and lizards
B forelimbs of bats and whales
C skeletons of turtles and crabs
D wings of locusts and blackbirds

24 The diagram that follows overleaf shows four of the stages that occur during the differentiation of phloem tissue in a plant. These occur in the order

A 2, 3, 1, 4. B 4, 1, 3, 2.
C 2, 3, 4, 1. D 3, 2, 1, 4.

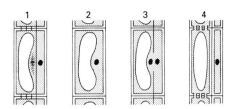

Items **25**, **26** and **27** refer to the following diagram which shows a possible arrangement of the genes involves in the induction of the enzyme β-galactosidase in the bacterium *Escherichia coli*.

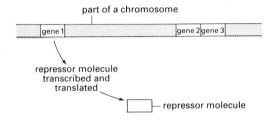

25 In the absence of lactose, the repressor molecule combines with gene 2 and, as a result, gene 3 remains 'switched off'. The correct identity of the three genes is

	operator gene	structural gene	regulator gene
A	2	1	3
B	1	3	2
C	3	2	1
D	2	3	1

26 In this system, the operon consists of
  A gene 2 only. B gene 3 only.
  C genes 2 and 3. D genes 1, 2 and 3.

27 Which of the following situations would arise if lactose became available to the cell?

	gene 1	gene 2	gene 3
A	−	−	+
B	−	+	+
C	+	−	+
D	+	+	+

(+ = 'switched on', − = 'switched off')

28 The following diagram shows an experiment set up to investigate abscission of leaf stalks.

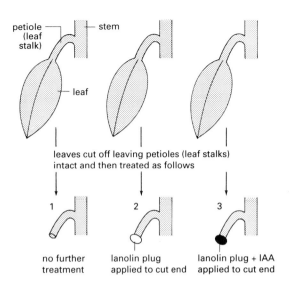

Following the treatment shown in the diagram, abscission of the leaf stalk would occur in
  A plant 1 only.   B plants 1 and 2.
  C plants 2 and 3. D plant 3 only.

29 Compared with the normal plant in the following diagram, which of the others show symptoms typical of nitrogen deficiency?

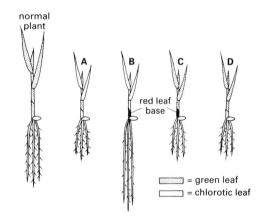

30 An individual organism's development depends on the interaction between its
  A genotype and environment.
  B environment and phenotype.
  C phenotype and chromosomes.
  D chromosomes and genotype.

31 An etiolated plant is characterised by possessing *both*
  A expanded yellow leaves and long strong internodes.
  B curled yellow leaves and long weak internodes.

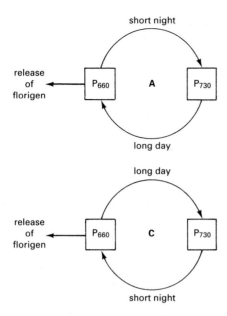

C expanded green leaves and short strong internodes.
D curled green leaves and short weak internodes.

32 Which of the diagrams at the top of the page correctly represents the events that occur during photoperiodism in a long day plant? (P = phytochrome)

33 The three numbered structures in the diagram below represent endocrine glands.

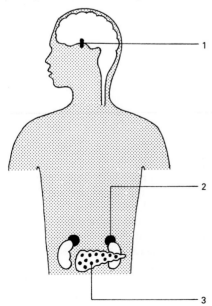

Which of the combinations in the following table correctly matches these glands with the hormones that they produce?

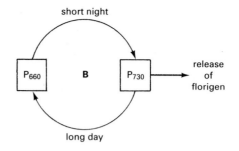

	endocrine gland		
	1	2	3
A	gonadotrophin	adrenaline	ADH
B	gonadotrophin	ADH	insulin
C	ADH	adrenaline	glucagon
D	ADH	glucagon	insulin

(ADH = antidiuretic hormone)

34 In the human body, which of the following does *not* occur in response to a sudden rise in environmental temperature?
A involuntary rhythmical contractions of skeletal muscles
B increase in blood circulation to body extremities
C increase in activity of sweat glands
D relaxation of erector muscles attached to hairs

35 Which of the following ecosystems would tend to remain most stable?

	relative state of ecosystem	predator-prey relationships
A	simple	only one prey species for each predator
B	complex	only one prey species for each predator
C	simple	many prey species for each predator
D	complex	many prey species for each predator

36 The table below refers to four species of whale.

species	estimated number before commercial whaling	estimated number today
Humpback	100 000	9 000
Fin	450 000	70 000
Sei	200 000	28 000
Right	50 000	3 000

Based on these estimates, which species of whale has undergone the greatest relative decrease in numbers?
A Humpback   **B** Fin
C Sei   **D** Right

37 The gills of a fresh water bony fish
A gain water by osmosis and absorb salts.
B lose water by osmosis and absorb salts.
C gain water by osmosis and excrete salts.
D lose water by osmosis and excrete salts.

38 It is best to transplant young seedlings on a day when the air is
A still, warm and dry.
B windy, warm and humid.
C still, cool and humid.
D windy, cool and dry.

39 In the table below, * indicates that the bird in the left column dominated the other bird by pecking it.

↓ bird against bird →	4	3	2	1
1	*	*		
2	*	*		*
3				
4		*		

The pecking order of the birds is
A 1, 2, 3, 4.   **B** 3, 4, 1, 2.
C 2, 1, 4, 3.   **D** 4, 3, 2, 1.

40 At the compensation point in a green plant
A photosynthesis comes to a halt.
B rate of respiration exceeds rate of photosynthesis.
C evolution of oxygen exceeds uptake of carbon dioxide.
D rate of synthesis of glucose equals rate at which it is used up.

ITEM NUMBER

	1	2	3	4	5	6	7	8	9	10	11	12	13	14	15	16	17	18	19	20
1																				
2																				
3																				
4																				
5																				
6																				
7																				
8																				
9																				
10																				
11																				
12																				
13																				
14																				
15																				
16																				
17																				
18																				
19																				
20																				
21																				
22																				
23																				
24																				
25																				
26																				
27																				
28																				
29																				
30																				
31																				
32																				
33																				
34																				
35																				
36																				
37																				
38																				
39																				
40																				

TEST NUMBER

SPECIMEN EXAMINATION

1	2	3	4	5	6	7	8	9	10	11	12	13	14	15	16	17	18	19	20

21	22	23	24	25	26	27	28	29	30	31	32	33	34	35	36	37	38	39	40